Tourism and Biopolitics in Pandemic Times

Maartje Roelofsen · Claudio Minca
Editors

Tourism and Biopolitics in Pandemic Times

palgrave
macmillan

Editors
Maartje Roelofsen ⓘ
Open University of Catalonia
Barcelona, Spain

Department of Economics
and Business
Universitat Oberta de Catalunya
Barcelona, Spain

Cultural Geography Group
Wageningen University and Research
Wageningen, The Netherlands

Claudio Minca ⓘ
Department of History and Cultures
University of Bologna
Bologna, Italy

ISBN 978-3-031-46398-3   ISBN 978-3-031-46399-0   (eBook)
https://doi.org/10.1007/978-3-031-46399-0

Cover credit: © Melisa Hasan

This Palgrave Macmillan imprint is published by the registered company Springer Nature
Switzerland AG
The registered company address is: Gewerbestrasse 11, 6330 Cham, Switzerland

Paper in this product is recyclable.

ACKNOWLEDGEMENTS

The idea for this book first came into being after a special session on 'Biopolitics and the Geographies of Tourism' at the 2022 Nordic Geographers Conference in Joensuu, Finland. After drafting the first version of this manuscript in the year that followed, we finally had the opportunity to meet the authors as contributors—digitally and in person—at a full-day workshop on 'Tourism and the Biopolitical', held at the Universitat Oberta de Catalunya (UOC) in Barcelona, Spain in May 2023. Without the financial support that Maartje Roelofsen received from the UOC to organize these events and to copy edit the book, this project would not have been possible. We therefore wish to express our sincere gratitude first and foremost to the Universitat Oberta de Catalunya and its Department of Economics and Business for their support, and also for funding Amos Ochieng's visit to the Barcelona workshop through the Research Connections Grant. We also want to thank the University of Bologna for funding Claudio Minca's and Caterina Ciarleglio's attendance to the Barcelona workshop. Finally, we wish to express our gratitude to Clare Patricia O'Sullivan, our copy editor, who did an excellent job and who has crucially contributed to the completion of the manuscript on time; thanks for your much appreciated dedication, Clare!

CONTENTS

List of Contributors

Christine Ampumuza Department of Tourism and Hospitality, Kabale University, Kabale, Uganda

Caterina Ciarleglio Department of History and Cultures, University of Bologna, Bologna, Italy

Myra Coulter Department of Urban and Tourism Studies, Université du Québec à Montréal, Montreal, QC, Canada

Dominic Lapointe Department of Urban and Tourism Studies, Université du Québec à Montréal, Montreal, QC, Canada

Chih-Chen Trista Lin Cultural Geography Group, Wageningen University and Research, Wageningen, Netherlands

Claudio Minca Department of History and Cultures, University of Bologna, Bologna, Italy

Amos Ochieng Department of Forestry, Biodiversity and Tourism, School of Forestry, Environmental and Geographical Sciences, Makerere University, Kampala, Uganda

Maartje Roelofsen Open University of Catalonia, Barcelona, Spain; Department of Economics and Business, Universitat Oberta de Catalunya, Barcelona, Spain; Cultural Geography Group, Wageningen University and Research, Wageningen, The Netherlands

LIST OF FIGURES

CHAPTER 1

Exploring Biopolitical Tourism Spatialities in Pandemic Times

Claudio Minca⊙ and Maartje Roelofsen⊙

Abstract This introduction chapter briefly discusses how biopolitics has been variously conceptualized in social theory and how biopolitics relates to tourism. Reflecting on the outbreak and development of the COVID-19 pandemic, the chapter also provides a critical review of the literature on the role of tourism as a key form of governance that impacts contemporary politics of mobility. The chapter concludes with an outline of the structure of the book.

C. Minca
Department of History and Cultures, University of Bologna, Bologna, Italy
e-mail: claudio.minca@unibo.it

M. Roelofsen (✉)
Open University of Catalonia, Barcelona, Spain
e-mail: mroelofsen@uoc.edu; maartje.roelofsen@wur.nl

Department of Economics and Business, Universitat Oberta de Catalunya, Barcelona, Spain

Cultural Geography Group, Wageningen University and Research, Wageningen, The Netherlands

© The Author(s), under exclusive license to Springer Nature Switzerland AG 2023
M. Roelofsen and C. Minca (eds.), *Tourism and Biopolitics in Pandemic Times*, https://doi.org/10.1007/978-3-031-46399-0_1

Keywords Biopolitics · Tourism · COVID-19 pandemic

In a highly regarded genealogical account of conceptualizations of the biopolitical, Thomas Lemke argues that biopolitics "places at the innermost core of politics that which usually lies at its limits, namely, the body and life. [Life] has become an abstraction, an object of scientific knowledge, administrative concern, and technical improvement" (2011, p. 117). Biopolitics, simply put, concerns all forms of politicization of life, including the management of populations, specifically in relation to questions of health and welfare, but also in relation to questions of mobility and the right to a place. This preoccupation with the qualification, quantification, and conceptualization of life is reflected in the various modes of management—on the part of state authorities and large corporations—of collective social bodies (and their formation and control) and, consequently, of individual bodies.

As far as mobility is concerned, biopolitical strategies have historically been employed to operationalize what Sheller (2018) has aptly identified as neoliberal versions of mobile (or relatively immobile) subjectivities, which are often conditioned by gender, race, and sexuality among other factors. While biopolitics refers more generally to processes of politicization of life, a biopolitical analytical approach may cast new light on the ways in which individual and collective mobility is conceptualized and managed by the authorities and on how this affects the life of the implicated populations. Given that tourism is viewed as a key manifestation of contemporary social and cultural life, and that the emergence of the so-called "new mobility paradigm" (ibid.) has crucially impacted our understanding of tourism in relation to the differential mobilities imposed on subjects by neoliberal regimes, it is intuitive to conclude that biopolitics is implicated in the production of present-day experiences of mass leisure travel and consumption.

The scholarship drawn together in this edited collection brings a biopolitical perspective to bear on selected tourism geographies associated with the COVID-19 pandemic (see, among others, Rose-Redwood et al., 2020). While the 'management of bodies' has always been a constitutive part of tourism and its spatialities (Alderman, 2018; Ek & Hultman, 2008; Minca, 2010, 2011; Veijola & Jokinen, 1994), the recent pandemic prompted the emergence of entirely new states of exception and emergency regimes, geared to exert tight restrictions and control over the mobility and embodied practices of millions of tourists and travellers. The vast and pervasive impact of COVID-19 made questions of mobility—including mobility 'for pleasure'—a core preoccupation for most governments across the globe, leading to the implementation of measures that had an enormous impact on the travel and tourism industry but also on contemporary understandings of the right to individual and collective mobility.

The circumstances of the pandemic have also pointed up the role of tourism as a key form of governance that impacts the politics of mobility (and the related freedom to move) and the associated selective practices within contemporary Western democracies. For example, the historical distinction between tourists and other 'people on the move', in terms of the right to cross borders and territories, became even more evident and geopolitically marked upon individual bodies during the global health crisis (see, among others, Cresswell, 2021; Dijstelbloem, 2021). Indeed, the right to travel was operationalized and disciplined using new forms of biometrics, bordering practices, and the collection and management of biomedical data, eventually to be incorporated into dedicated documents, such as the European Union (EU) *Digital COVID Certificate*, the *IATA Travel Pass*, the *NHS COVID Pass*, and other regional variations. Between 2021 and 2023, these and other internationally recognized biopolitical applications assisted the decentralized verification of interoperable vaccine-, testing-, and recovery certificates, thereby facilitating the mobility of selected categories of individuals between places, regions, and countries. Specific technologies enabling the rapid and easy checking of body temperature were developed and universally deployed to monitor, record, restrict, and facilitate the mobility of tourists as newly conceived biological populations. Many governments and the tourism industry heralded the adoption of these (digital) tools as instrumental to an efficient response to the economic crisis provoked by the pandemic (UNWTO, 2021). While such forms of biometric monitoring were employed first and foremost to address and mitigate the health emergency, they were ultimately also enforced to enable tourists to travel under supposedly safer conditions, thereby contributing to the recovery of local, regional, and national tourism economies.

Critical voices have pointed out the potentially divisive and discriminatory effects of biopolitical initiatives impinging upon the mobility of selected groups of individuals. Getting vaccinated, tested, or receiving treatment to recover from a COVID-19 infection crucially relies on the availability *of* and access *to* health care services, which varied considerably across countries and even across regions during the crisis (Bialasiewicz &

Alemanno, 2021). Furthermore, not all social groups have equal access to such services: those who were already marginalized in terms of their access to healthcare before the pandemic have been the most disadvantaged by the new circumstances, particularly in non-white, non-national, and other disadvantaged communities (Whitacre et al., 2021). Similarly, those whose lives were prioritized in terms of earlier access to vaccines, such as individuals with underlying health conditions or ageing populations—the so-called 'high-risk' groups—were differently defined across countries, leading to asymmetrical categorizations of those worthy of protection and those not. In addition, the cost of getting tested for SARS-CoV-2 and receiving certificates authorizing mobility through spaces and across borders (particularly for non-essential purposes such as leisure travel) represented a substantial barrier to lower-income groups, especially in countries without a universal healthcare system (ibid.). For these groups, the lack of resources available to invest in the prevention of illness and the promotion of biological wellbeing reflected the fact that "the valuation of life is materially embedded in economic relations", including those relations produced through tourism (Krupar & Ehlers, 2020, p. 441). Clearly, then, the pandemic measures were implemented to foster some lives and related tourism mobilities, whereas others were left prone to infection, illness, and death.

While new configurations of tourist mobilities were gradually sanctioned *and* facilitated by means of the interventions we have described, at the same time, the perception of a right-to-go-on-vacation incentivized the invention of creative yet questionable new experimental tourist practices, including the emergence of supposedly "COVID-free" tourist spaces. These include COVID-free cruises and COVID-free islands (see Chapter 2 in this volume) or the presumably safe 'pilot holidays' promoted by the Dutch government during the pandemic (see Minca & Roelofsen, 2023). Such experimental biopolitical tourist laboratories were mainly informed by the neoliberal approaches to the health crisis of some governments, especially those that were famously inspired by herd immunity fantasies (and strategies). These and other initiatives to incentivize tourism (but not other) mobilities during the health crisis were mainly designed by the industry, in collaboration with certain governments, to keep the (tourist) 'show going on' despite the unprecedented threat represented by COVID-19, and provoked what remains a fundamentally unmonitored harm to the populations under the implicated governments' tutelage (Roelofsen & Cássia Ariza da Cruz, 2023). As noted by Roberto Esposito—a key figure in contemporary debates surrounding questions of immunity and immunization—the implementation of strategies developed around the notion of herd immunity (and the associated management of individual mobility and social contact) amounted to a form of *thanatopolitics* based on the deliberate sacrifice (and greater exposure to illness) of the most vulnerable and 'less fit' members of the population in favour of free circulation and a more active life for the most productive members (Esposito, 2023; Minca & Roelofsen, 2023).

Over the past few years, a rich and diverse body of academic work has emerged in response to the pandemic-induced crisis of leisure travel (see, among many others, Brouder et al., 2020; Lew et al., 2020; Mostafanezhad et al., 2020). The literature reviewed in the next chapter (see Ciarleglio & Minca, this volume) suggests that research on the tourism-related impact of the COVID-19 emergency has largely focused on the leading consequences of the crisis for tourism workers and on strategies related to vaccination campaigns and recovery plans implemented by public authorities and the private sector. A further key segment of the debate has concerned broader questions of mobility and immobility, and the ways in which the pandemic has further accentuated the selective nature of contemporary travel, in terms of the different fields of possibility afforded to different subjects depending on their social and economic status, but also on factors such as race, gender, and sexuality. Relatedly, this recent global health crisis has prompted discussion surrounding the new forms of regulation of bodily conduct and the associated practices of surveillance implemented by public authorities and tourism industry (see Lapointe & Coulter, 2020). The global health emergency has also been examined in relation to other global crises affecting the present and the future of tourism, including the environmental crisis associated with climate change, the migration crisis, and the rising tensions in global politics (for a critique, from a biopolitical perspective, see Tzanelli, 2022).

Now more than ever, questions regarding individual and collective health and wellbeing have been propelled to the centre of public debate, not least because during the COVID-19 pandemic, tourism provided in many instances the socio-spatial conditions for the virus to spread. For example, while tourism infrastructures such as hotels and cruise ships functioned as vectors for the virus, they also become essential spaces for quarantine, containment, and 'human behavioural experiments' (Minca & Roelofsen, 2023). Nevertheless, despite growing interest in the 'politics of life' within tourism studies, only limited attention has been paid to the biopolitical aspects of the pandemic in the context of tourism (an exception being Iaquinto, 2020; Iaquinto et al., 2023). Indeed, the so-called 'biopolitical turn' in the humanities and the social sciences has only been partially assimilated by the tourism literature. According to Lapointe and Coulter (2020), who recently offered a comprehensive and compelling review of this specific body of work, most biopolitical analyses of tourism rely on the original framing of biopolitics advanced by Michel Foucault (e.g., Ek & Hultman, 2008; Lunden, 2022; Maurette & Fguira, 2022). Some authors have also drawn on Giorgio Agamben's speculations on the biopolitical in relation to sovereign power and the state of exception (e.g., Diken & Laustsen, 2004; Minca, 2011; Renaud, 2022). Additionally, recent research has been informed by the work of other Italian political philosophers, such as Roberto Esposito's reading of biopolitics as a manifestation of what he calls the immunization paradigm affecting all modern Western political systems (e.g., Minca, 2011; Roelofsen & Minca, 2018; Minca & Roelofsen, 2023), or Michael Hardt and Antonio Negri's critique of biocapitalism and its affirmative potential (e.g., Constable, 2016; Simpson, 2016). Affirmative biopolitics has also been discussed in the tourism literature with respect to Rosi Braidotti's and Lauren Berlant's feminist perspectives on the politicization of the bios (Lin & Minca, 2020; Lin et al., 2018). Overall, this body of work has interrogated both the relationship between biopower and questions of race, gender, disability, and sexuality, and the scale of broader tourism business strategies and the related notion of tourist experience. Given that the industry has displayed a continued strategic focus on how to represent and concretely manage the tourist body—in this regard often appealing to notions of happiness and wellbeing—the critical literature has brought corresponding scrutiny to bear on the spatio-temporal regimes of enclavic tourism spaces, and on the associated promotion (and protection) of specific bodies—together with the exclusion of others. Also inherently implicated in this line of inquiry is the representation of, and the performances and roles played by, the bodies of enclavic tourism workers, who are often presented as desubjectified smiling beings, as part of a landscaped environment, and part of the overall tourist product and experience (Minca, 2010). Meanwhile, biopolitical perspectives have also been applied to other tourist spaces. A very recent study (Iaquinto et al., 2023), for example, devoted particular attention to the exercise of biopower in art museums, resort hotels, casinos, and airplanes, all places frequented and used by tourists. Each of these sites reflects the somewhat coercive power of tourism, which draws on myriad spatial technologies of safety and surveillance to produce healthy, happy, and docile citizens and consumers, but usually at the expense of those deemed to be of lesser value, including tourism workers (see also Sheller, 2021).

A recent special issue on tourism and biopolitics in the journal *VIA Tourism Review* teased out the different intersecting biopolitical dimensions of tourism (Lapointe, 2022; see also Lapointe & Coulter, 2020), including *technologies* with the capacity to orchestrate, control, and evaluate tourism services and interactions. Accommodation platforms like Couchsurfing and Airbnb are examples of such technologies; in bringing together hosts and guests, they produce economies of affect, but at the same time users often experience their networking capabilities as disciplinary and oppressive (Roelofsen & Minca, 2018). These platforms engage in a de facto form of biopolitical governance: in particular, they dictate a specific set of standards to which the practices of hosts and guests are expected to conform, via myriad embodied performances in intimate home environments (ibid.). Yet, in other instances, they are appropriated for connective action and emancipation, particularly among touring subjects whose bodies have endured racial and gendered discrimination in their search for hospitality (Costa, 2022). For such individuals, accommodation platforms can act as key tools for establishing travelling communities centred around shared political ideologies. A further biopolitical dimension of tourism identified in the *VIA* special issue is *placemaking*, understood as a spatial strategy that simplifies, essentializes, and 'makes and breaks' spaces. The Arctic, for example, has been framed as a place "for the purposes of touristic consumption through a spectrum of planetary care" and sustainability, from a homogenized perspective that is extended to the whole of Northern Europe (Lunden, 2022, p. 216). Again, in the special issue, a case study on Bokong Nature Reserve in Lesotho illustrates how certain rural areas have been strategically re-territorialized and bordered to protect selected ecosystems. These are then offered up for tourism consumption, but at the expense of local communities who are evicted and prohibited from using the protected spaces (Mokhele, 2022). Studies such as these cast into stark relief the grimmer faces of tourism and the injustices that it can produce as a biopolitical force. Hence, Tzanelli responds by providing tourism scholars with helpful pointers on how to incorporate biopolitical analysis into their future work, especially in terms of exploring "the pragmatic aspects of human agency over power and corporate violence, discarding the idea of complete human automation/subjection" (2022, p. 951). This may be done by investigating real-life instances of biopolitics and biopower operating through tourism, as well as examples of pragmatic action driving social change in tourism and hospitality contexts that might be understood as counter-biopolitical (ibid.). Tzanelli further argues that scholars should discard practices that are seemingly driven by a commitment to justice, but which de facto "marginalise[…] the 'objects' of care, such as exploited labour, indigenous populations and natural ecosystems acting as tourist destinations" (ibid.).

While we share most of the perspectives and critical considerations put forward in this literature, we nevertheless believe that applying a biopolitical lens to the realm of tourism raises key methodological questions. With few exceptions, a large part of the existing work comprises theoretical contributions that draw on real-world examples in support of their main arguments. As compelling as such theorizing may be, further work is required—in our view—to place the biopolitical analysis of empirical cases on firmer ground via methodologies suited to addressing the major challenges posed by the 'biopolitical in tourism'. Furthermore, the debate on the biopolitical is now extremely rich and diverse and is consequently marked by tension between different positionings (in relation to geography, see, for example, Minca et al., 2022). How do different articulations and interpretations of biopolitics (for example the so-called 'negative versus affirmative' biopolitics debate) affect the possibility of implementing methodologies for unpacking the empirical implications of a biopolitical strategy? How might such divergent perspectives condition the field of tourism? Thus, the concept for this volume stemmed from our desire to begin empirically investigating selected forms of the biopolitical in tourism during pandemic times. Needless to remark, this edited collection represents an incomplete, embryonic attempt to engage with such an enormous and ambitious task. Nevertheless, we have aimed to provide a stimulating start by inviting the input of tourism scholars who bring different geographical perspectives to bear on the biopolitical in tourism, while drawing on compelling empirical cases. Across its discussion of case studies on the unfolding of the pandemic across a variety of contexts, the collection revisits classical biopolitical approaches to tourism (Chapters 2 and 3), but also contemplates more-than-human biopolitical approaches (Chapter 4), and affirmative biopolitical approaches (Chapter 5). Our overarching purpose here is to sort through how a biopolitical lens may contribute to the analysis of different practices and regimes of mobility, security, and in/exclusion of specific individuals, bodies, and populations in the context of tourism.

Following in the tradition of biopolitical analysis within tourism studies, Caterina Ciarleglio and Claudio Minca in Chapter 2 examine an experiment with a 'COVID-free' island, which was promoted by the Italian government as an immunized enclavic space at a certain stage of the pandemic. Islands, precisely due to their perceived separateness from 'the rest of the world', have often been presented as protected environments and as ideal settings for the materialization of 'utopian' spaces where ad hoc tourist communities may be re-created and the individual bodies composing them may be carefully managed (Minca, 2010). Thus, the very notion of an experimental tourist community situated in an enclavic space on an island draws on a long tradition of tourist 'laboratories' where tourist bodies may be closely governed and monitored (Ek & Tesfahuney, 2019; Simpson, 2016). At a time when 'COVID-free' spaces and travel corridors were becoming a putative new horizon to be explored by relevant industry actors, some islands were advertised as potentially shielded from infection and ready to engage in close and continuous monitoring of the individual and collective behaviours and movements of 'guests'. However, no enclavic tourist space is entirely... 'insulated'. Indeed, as shown in the case study on the island of Favignana presented in Chapter 2, the attempt to promote this destination as an immunized, COVID-free space resulted in the arrival of an almost entirely unmonitored mass of tourists in the summer of 2021. These tourists, furthermore, displayed poor compliance with protective measures, thus frequently exposing tourism workers to the risk of infection. The chapter also discusses the narratives that accompanied the concept of 'COVID-free' tourist spaces in Italy and their fruition, while also considering how these narratives were operationalized by the authorities via dedicated vaccination campaigns for residents and specific spatial strategies related to the mobility of the tourists.

In Chapter 3, Myra Coulter and Dominic Lapointe present a discourse analysis of articles published by international English-language news outlets, in relation to the publicization and politicization of the embodied and affective experiences of tourism and travel industry stakeholders during the earliest months of the pandemic. Tracing the unfolding of the global health crisis, the authors focus on a specific embryonic moment in the evolution of the global tourism industry, applying a biopolitical lens to the narratives of uneven (im)mobile subjects as represented in the mass media. Rather than a renegotiation of privileged mobilities in a world grappling with COVID-19, as imagined by many critical and regenerative tourism scholars and tourist industry stakeholders, this chapter reveals a chaotic reordering of normative social, spatial, and mobile hierarchies. While intended to protect public health, this reordering led to the control of leisure mobilities and, by extension, to the collapse of the tourism economy. The bodies of global travellers not only came under threat from the novel coronavirus but also served as vehicles for it, spreading disease within and beyond leisure spaces (Adey et al., 2021) and triggering health alerts for many people. Hence, Chapter 3 illustrates the uneven representations of variously positioned (im)mobile subjects, including the amplification of tourist experiences gone awry and the relative silencing of worker and resident voices. It thereby documents the reinforcement of prior biopolitical arrangements of subjectivities and spatial (im)mobilities in Western societies (Rose, 2021).

In Chapter 4, Amos Ochieng, Christine Ampumuza, and Maartje Roelofsen push beyond anthropocentric understandings of biopolitics and reflect on how the biopolitical operates on/with more-than-human lives in tourism. They highlight some of the bio- and necropolitical dimensions of conservation-based tourism and discuss the impact of the pandemic in the context of gorilla tourism at Bwindi Impenetrable National Park in Uganda. Gorilla tourism has typically been implemented and organized around 'habituation', a construct often defined within conservation and tourism studies as a relational process through which animals and humans become accustomed to one another's presence. Among other purposes, habituation allows animals to become 'viewable' as subjects/objects for tourism consumption and scientific study, which arguably helps local communities in terms of the related economic benefits. While habituation processes may be guided by the desire to exert human biopolitical control over nature for human satisfaction, nonhuman animals such as the Bwindi gorillas took on an active role in shaping a 'livelier' and perhaps more affirmative biopolitics during the pandemic, producing surprising and unforeseen relations. Chapter 4 considers the implications of the pandemic-related halting of tourism mobility for gorilla-tourist interaction and habituation in Bwindi. In so doing, it draws upon and contributes to posthuman and more-than-human biopolitical theory and relational approaches to animal agency in tourism.

Finally, in Chapter 5, Trista Chih-Chen Lin (conceptually and politically) re-imagines the biopolitical by foregrounding forms of affirmative ethics (Braidotti, 2016) and affirmative biopolitics. Lin focuses on a special category of tourist workers—flight attendants—who, during the early stages of the pandemic, were extensively exposed to risk of infection by the virus by virtue of their in-flight duties. She offers an analytical framework for 'affirmative alternatives' to the pandemic biopolitics normally applied to tourist workers, drawing inspiration from the work of Lauren Berlant, Rosi Braidotti, Michael Hardt, Antonio Negri, and Soile Veijola. Using the Taoyuan Flight Attendants Union (TFAU) in Taiwan as a case study, the author explores three types of campaigns for infection control and occupational health led by the union between 2020 and 2022. The chapter highlights the importance of organized immaterial and affective labour, not only as an object of (ongoing) governmental control and corporate exploitation (during and beyond the pandemic), but also as a potential source of, and collective capacity for, a new and empowering form of sociality within tourism work. This would be an alternative sociality to that previously found in air travel and to that structured by the regime of disease control. By way of conclusion, Lin argues for the importance of tracking subjectivity and relational affectivity when critiquing and identifying life-affirming alternatives to the past and present regulation of health, leisure, and hospitality in air travel, tourism, and beyond.

It is thus now time to initiate our journey around these different international locations, where we shall explore some of the empirical implications of the biopolitical in tourism in times of COVID-19, while bearing in mind the utterly experimental and partial nature of this effort. We hope that readers will appreciate how 'grounding' different strands of biopolitical analysis in different contexts may stimulate further inquiry in this direction, while simultaneously confirming the key role of biopolitics in contemporary tourism, a role that the recent pandemic has helped to disclose in all its relevance.

Competing Interests

The authors have no conflicts of interest to declare that are relevant to the content of this chapter.

References

Adey, P., Hannam, K., Sheller, M., & Tyfield, D. (2021). Pandemic (Im)mobilities. *Mobilities, 16*(1), 1–19. https://doi.org/10.1080/174 50101.2021.1872871

Alderman, D. H. (2018). The racialized and violent biopolitics of mobility in the USA: An agenda for tourism geographies. *Tourism Geographies, 20*(4), 717–720. https://doi.org/10.1080/14616688.2018.1477168

Bialasiewicz, L., & Alemanno, A. (2021). *The dangerous illusions of an EU 'vaccine passport.'* Open Democracy. https://www.opendemocracy.net/en/can-europe-make-it/the-dangerous-illusions-of-an-eu-vaccine-passport/

Braidotti, R. (2016). Posthuman affirmative politics. In S. E. Wilmer & A. Zukauskaite (Eds.), *Resisting biopolitics: Philosophical, political, and performative strategies.* (pp. 30–56). Routledge.

Brouder, P. (2020). Reset redux: Possible evolutionary pathways towards the transformation of tourism in a COVID-19 world. *Tourism Geographies, 22*(3), 484–490. https://doi.org/10.1080/14616688.2020.1760928

Brouder, S. T., Salazar, B., Mostafanezhad, M., Mei Pung, J., Lapointe, D., Higgins Desbiolles, F., Haywood, M., Hall, C. M., & Balslev Clausen, H. (2020). Reflections and discussions: Tourism matters in the new normal post COVID-19. *Tourism Geographies, 22*(3), 735–746. https://doi.org/10.1080/14616688.2020.1770325

Constable, N. (2016). Reproductive labor at the intersection of three intimate industries: Domestic work, sex tourism, and adoption. *Positions: Asia Critique, 24*(1), 45–69. https://doi.org/10.1215/10679847-3320041

Costa, T. (2022). Travel politicization facing biopolitics: (Im)mobilities, technological mediations, and identity discourses in collaborative hosting networks for women. *Via Tourism Review, 21,*. https://doi.org/10.4000/viatourism.8353

Cresswell, T. (2021). Valuing mobility in a post COVID-19 world. *Mobilities, 16*(1), 51–65. https://doi.org/10.1080/17450101.2020.1863550

Diken, B., & Laustsen, C. B. (2004). Sea, sun, sex and the discontents of pleasure. *Tourist Studies, 4*(2), 99–114. https://doi.org/10.1177/146879760 4054376

Dijstelbloem, H. O. (2021). *Borders as infrastructure: The technopolitics of border control.* The MIT Press.

Ek, R. (2016). The tourist camp: All-inclusive tourism, hedonism and biopolitics. In M. Tesfahunay & K. Schough (Eds.), *Tourism as world ordering* (pp. 47–66). Cambridge Scholars Publishing.

Ek, R., & Hultman, J. (2008). Sticky landscapes and smooth experiences: The biopower of tourism mobilities in the Öresund region. *Mobilities, 3*(2), 223–242. https://doi.org/10.1080/17450100802095312

Ek, R., & Tesfahuney, M. (2019). Topologies of tourism enclaves. *Tourism Geographies, 21*(5), 864–880. https://doi.org/10.1080/14616688.2019.1663910

Esposito, R. (2023). *Common immunity. Biopolitics in the age of the pandemic.* Polity Press.

Iaquinto, B. L. (2020). Tourist as vector: Viral mobilities of COVID-19. *Dialogues in Human Geography, 10*(2), 174–177. https://doi.org/10.1177/2043820620934250

Iaquinto, B. L., Cheer, J. M., & Roelofsen, M. (2023). Coercive geographies: Biopower, spatial politics, and the tourist. *Environment and Planning C: Politics and Space.* https://doi.org/10.1177/23996544231194828

Krupar, S., & Ehlers, N. (2020). Biocultures: A critical approach to mundane biomedical governance. *Culture, Theory and Critique, 61*(4), 440–456. https://doi.org/10.1080/14735784.2020.1857810

Lapointe, D. (2022). Apparatus for control, state of exception and tourist bodies: Tourism as a biopolitical phenomenon. *Via Tourism Review, 21.* https://doi.org/10.4000/viatourism.8070

Lapointe, D., & Coulter, M. (2020). Place, labor, and (Im)mobilities: Tourism and biopolitics. *Tourism Culture & Communication, 20*(2), 95–105. https://doi.org/10.3727/109830420X15894802540160

Lemke, T. (2010). *Biopolitics.* NYU Press.

Lew, A. A., Cheer, J. M., Haywood, M., Brouder, P., & Salazar, N. B. (2020). Visions of travel and tourism after the global COVID-19 transformation of 2020. *Tourism Geographies, 22*(3), 455–466. https://doi.org/10.1080/14616688.2020.1770326

Lin, C.-C.T., Minca, C., & Ormond, M. (2018). Affirmative biopolitics: Social and vocational education for Quechua girls in the postcolonial 'affectsphere' of Cusco, Peru. *Environment and Planning d: Society and Space, 36*(5), 885–904. https://doi.org/10.1177/0263775817753843

Lin, C.-C.T., & Minca, C. (2020). Affective life, 'vulnerable' youths, and international volunteering in a residential care programme in Cusco, Peru. *Transactions of the Institute of British Geographers, 45*(4), 735–749. https://doi.org/10.1111/tran.12374

Lunden, A. (2022). The biopolitics of Artic tourism development and sustainability. *Via Tourism Review, 21.* https://doi.org/10.4000/viatourism.8084

Maurette, T., & Ben Fguira, S. (2022). A case of medical tourism? *Via Tourism Review, 21,.* https://doi.org/10.4000/viatourism.8590

Minca, C. (2010). The Island: Work, tourism and the biopolitical. *Tourist Studies, 9*(2), 88–108. https://doi.org/10.1177/1468797609360599

Minca, C. (2011). No country for old men. In C. Minca & T. Oakes (Eds.), *Real tourism: Practice, care, and politics in contemporary travel culture* (pp. 12–37). Routledge. https://doi.org/10.4324/9780203180969-8

Minca, C., Rijke, A., Pallister-Wilkins, P., Tazzioli, M., Vigneswaran, D., van Houtum, H., & van Uden, A. (2022). Rethinking the biopolitical: Borders, refugees, mobilities.... *Environment and Planning C: Politics and Space, 40*(1), 3–30. https://doi.org/10.1177/2399654420981389

Minca, C., & Roelofsen, M. (2023). Post-COVID biopolitical fantasies and the case of the Dutch 'Pilot Holidays.' *Environment and Planning C: Politics and Space.* https://doi.org/10.1177/23996544231194828a

Mokhele, I. (2022). The biopolitical production of tourism areas in Lesotho: The case of Bokong Nature Reserve. *Via Tourism Review, 21.* https://doi.org/10.4000/viatourism.8528

Mostafanezhad, M., Cheer, J. M., & Sin, H. L. (2020). Geopolitical anxieties of tourism: (Im)mobilities of the COVID-19 pandemic. *Dialogues in Human Geography, 10*(2), 182–186. https://doi.org/10.1177/2043820620934206

Renaud, L. (2022). Strategy for controlling tourist mobility: Analysis of the biopolitical processes of territorialization implemented by the cruise tourism industry in a Caribbean destination. *Via Tourism Review, 21,*. https://doi.org/10.4000/viatourism.8254

Rose-Redwood, R., Kitchin, R., Apostolopoulou, E., Rickards, L., Blackman, T., Crampton, J., Rossi, U., & Buckley, M. (2020). Geographies of the COVID-19 pandemic. *Dialogues in Human Geography, 10*(2), 97–106. https://doi.org/10.1177/2043820620936050

Roelofsen, M., & Cássia Ariza da Cruz, R. (2023). Denialist and neoliberal approaches to tourism and the COVID-19 pandemic. *Tourism Geographies.* https://doi.org/10.1080/14616688.2023.2269546

Roelofsen, M., & Minca, C. (2018). The superhost. Biopolitics, home and community in the Airbnb dream-world of global hospitality. *Geoforum, 91,* 170–181. https://doi.org/10.1016/j.geoforum.2018.02.021

Rose, J. (2021). Biopolitics, essential labor, and the political-economic crises of COVID-19. *Leisure Sciences, 43*(1–2), 211–217. https://doi.org/10.1080/01490400.2020.1774004

Sheller, M. (2018). *Mobility justice: The politics of movement in an age of extremes.* Verso.

Sheller, M. (2021). Mobility justice and the return of tourism after the pandemic. *Mondes du tourisme,* https://journals.openedition.org/tourisme/3463

Simpson, T. (2016). Tourist utopias: Biopolitics and the genealogy of the post-world tourist city. *Current Issues in Tourism, 19*(1), 27–59. https://doi.org/10.1080/13683500.2015.1005579

Tzanelli, R. (2022). Biopolitics in critical tourism theory: A radical critique of critique. *Via. Tourism Review, 21.* http://journals.openedition.org/viatourism/8242

United Nations World Tourism Organization. (2021). *COVID-19: Measures to support travel and tourism.* https://www.unwto.org/tourism-data/covid-19-measures-to-support-travel-tourism

Veijola, S., & Jokinen, E. (1994). The body in tourism. *Theory, Culture & Society, 11*(3), 125–151. https://doi.org/10.1177/026327694011003006

Whitacre, R., Oni-Orisan, A., Gaber, N., Martinez, C., Buchbinder, L., Herd, D., Holmes, M., & S. (2021). COVID-19 and the political geography of racialisation: Ethnographic cases in San Francisco, Los Angeles and Detroit. *Global Public Health, 16*(8–9), 1396–1410. https://doi.org/10.1080/17441692.2021.1908395

"A Healthy Person is a Happy Person". Biopolitical Reflections on the Promotion of Favignana as a COVID-free Island

Caterina Ciarleglio and *Claudio Minca*

Abstract This chapter discusses the experiment of 'COVID-free' island, promoted as an enclavic immunised space by the Italian government during the pandemic. The idea of an experimental tourist community placed in an enclavic space located on an island draws from a long tradition of tourist 'laboratories' in which tourist bodies may be closely governed and monitored. In a time in which 'COVID-free' spaces and travel corridors have become a new putative horizon to be explored by the relevant industry, some islands have been advertised as potentially shielded from contamination and available for close and continuous monitoring of individual and collective behavior of 'the guests' and their movements. As the analysis of the island of Favignana developed in this chapter

C. Ciarleglio (✉) · C. Minca
Department of History and Cultures, University of Bologna, Bologna, Italy
e-mail: caterina.ciarleglio@studio.unibo.it

C. Minca
e-mail: claudio.minca@unibo.it

17

M. Roelofsen and C. Minca (eds.), *Tourism and Biopolitics in Pandemic Times*, https://doi.org/10.1007/978-3-031-46399-0_2

demonstrates, the attempt to promote this destination as an immunized COVID-free space has resulted in the arrival of an almost unmonitored mass of tourists in the summer 2021, in a relatively poor respect of the protective measures and in the exposure of the tourist workers to the frequent possibility of been contaminated.

Keywords COVID-free islands · Biopolitics · Tourist enclaves · Esposito · Paradigm of immunisation · Favignana

"A healthy person is a happy person". This bold statement strikingly pervades the opening scenes of "The Island" (Parkers, 2005), a film that we already drew on some years ago in an initial attempt at conceptualising the relationship between tourism and biopolitics (Minca, 2010). Indeed, "The Island" portrays a sort of tourist resort where human clones are produced—unbeknownst to themselves—under the strict control of a corporation, as a means of generating healthy (and thus potentially happy) bodies whose organs will be harvested to replace those of wealthy clients interested in curing illnesses or prolonging their lives. Hence, the slogan "A healthy person is a happy person" is continuously announced over the public address system—against a mildly Orwellian backdrop— to remind the 'guests' to eat and exercise appropriately, and to respect, for the sake of their own well-being, the pace of life imposed by the resort's spatio-temporal regime. Arguably, the resort in the movie bears multiple resemblances to the typical 'enclavic tourist space', which is marked by panoptic control and a rigorous spatio-temporal framework for both guests and hosts/workers. Later in the chapter, we shall return to the question of enclavic tourist spaces in relation to biopolitics. For now, suffice it to say that the slogan (and the movie overall) provocatively reflects what might be viewed as a constitutive element of many tourist geographies: that is to say, the leading role of tourist bodies and, accordingly, the importance of managing the care and control of these bodies, especially when they are confined upon dedicated tourist 'islands'.

The emergence of the SARS-CoV-2 pandemic, and the related implications for the mobility of tourists across the globe, has further highlighted the importance to the tourist industry of managing bodies. The measures adopted by many governments, beginning in early 2020, to contain the spread of the virus and, at the same time, to protect and boost the tourism economy, are provocatively evoked by the slogan cited above. Among such measures, some European governments began experimenting with 'COVID-free islands'. More specifically, they attempted to create totally immune spaces where, thanks to the blanket vaccination of residents, tourists could behave (and feel) as though the pandemic did not pose any threat to them. In relation to these experimental biological laboratories, set up with the specific aim of selectively supporting the tourist industry during the pandemic, our opening statement might be reformulated as "a happy person *must necessarily be* a healthy person". With a view to unpacking the intimate connection between biopolitics and the management of the tourist body in the COVID-19 era, we set out here to explore how the Italian island of Favignana, which is part of the Aegadian Islands archipelago off the coast of Sicily, was promoted as a 'COVID-free island' beginning in June 2021. Based on textual analysis and ethnographic work conducted during the summer of 2021, we discuss how the desire of the central government and local authorities to incentivise tourism recovery on the island led to the prioritisation of vaccinating the local population (compared, for example, to other, more vulnerable groups living in other parts of the country), as well as to a new kind of immunitarian geography associated with the biopolitical management of the bodies of tourists, workers, and residents.

The aim of this chapter is to contribute to existing debates on the relationship between tourism and the pandemic by drawing on a biopolitical perspective informed by Roberto Esposito's conceptualisation of the "immunization paradigm" (Esposito & Campbell, 2006) as a key driver within modern Western politics. To this end, we first outline the COVID-free island experiment more broadly, and then move on to offer a detailed analysis of how Esposito's analytical framework of immunisation can shed light on the biopolitical dimension of this kind of experimentation (see, for example: Ajana, 2021). Thus, in the following sections, we focus on the narratives associated with the promotion of Favignana as an immunised space and show that tourism may be seen as a powerful social, cultural, and economic transformative force, especially in an insular context such as that under study here. We conclude with some final considerations on the importance of a biopolitical reading of tourism, and how this may be related to broader questions of health, privilege, and the spatialisation of capitalist accumulation via leisure consumption.

"Dreaming of Immunized Utopias": The COVID-free Islands Project as a Biopolitical Experiment

The literature on the impact of the COVID-19 pandemic on tourism over the 2020–2021 period—when measures to contain the spread of the virus were being implemented by governments across the globe—may tentatively be seen as clustered around two bodies of work. First, there is the 'tourist management' literature, which has largely focused on the management and control of the virus (and of tourists), potential recovery strategies for the tourism sector as a whole, risk perception, and comparisons with previous pandemics. Second, there is work in the broader field of 'tourism studies', which has mainly offered critical theoretical interventions on the broader impact of the pandemic on global tourism, often presenting the emergency as an opportunity to explore new forms of tourist mobility informed by principles of degrowth and sustainability.

More specifically, debates surrounding the first cluster of studies have emphasised a series of practical implications associated with the pandemic, including questions of financial support for tourism workers previously employed without regular contracts (Connell, 2021; Khalid et al., 2021; Williams, 2021) and the differential impact of *immobility* for tourism workers, reflecting widespread inequalities in terms of legal rights and access to support in times of crisis (Aronica et al., 2021; Lin et al., 2021; Bichler et al., 2021). Other work has highlighted the strategic importance of vaccination campaigns (Zhu et al., 2022), and the role played by travel insurance companies and by media communications surrounding risks to travellers (Cheng et al., 2022; Tan & Caponecchia, 2021; Gu et al., 2021), as well as the strategic timing of interventions in support of the sector and the need for alternative plans and strategies for the future (Kowalewski, 2021; Nuna & Paulo, 2020; Samdin et al., 2022).

The second body of work has largely explored broader questions of mobility in light of the pandemic (Cresswell, 2020; Jensen, 2021; Salazar, 2021) in an attempt to propose a radical revisiting of tourism and mass mobility (Haywood, 2020; Higgins-Desbiolles, 2020; Hall et al., 2020; Lew et al., 2021)—especially in relation to various manifestations of the present social crisis (Hopkins, 2021), political crisis (Oakes, 2020), financial crisis (Butcher, 2021), or broader environmental crisis (Ioannides et al., 2020; Mostafanezhad, 2020). A similar set of issues has also proved salient during the final stages of the pandemic. With the lifting of containment measures, the related health crisis has continued to be presented in this literature either as an opportunity to reimagine the future development of tourism, whether in the urban context (Paddison & Hall, 2022) or in virtual spaces (Barker & Rodway-Dyer, 2022; Dybsand, 2022), or as a tool for advancing more environmentally conscious forms of mobility (Garcia et al., 2022; Gössling & Schweiggart, 2022; Nikolaeva et al., 2023). Another line of inquiry has offered critical reflections on the impact of the pandemic on the hospitality industry as a whole (Brune et al., 2023; Xiang et al., 2022), on broader questions of socially selected immobility (Wanchan & Haines, 2022), and on the associated management strategies (Castañeda-Garcia et al., 2022; Husser et al., 2023).

Arguably, while the existing literature has discussed the impact of the pandemic on small tourist islands (Connell, 2021; Gu et al., 2021), no work has yet considered the cases of the COVID-free islands experimentally promoted by some European governments during the second phase of the pandemic. The first COVID-free islands were conceptualised (and implemented) by the Greek government in spring 2021, followed by the Italian government a few months later (*The Guardian*, 6/04/2021). The Greek executive's plan was specifically designed to render certain islands 'immune' from the virus, with a view to avoiding—or at least mitigating—the impact on the tourist industry of the restrictive measures that had been imposed globally on individual mobility the previous year. The tourist sector is essential to the Greek economy (generating about one fourth of the country's GDP and employing about 15% of its labour force) (*Le Figaro*, 11/5/2021). Due to the pandemic, over 60,000 jobs were lost in the sector in 2020, an impact that was especially dramatically felt on the small islands in the Aegean Sea, where tourism represents the main source of income and jobs (*Euronews*, 14/5/2021). The declared objective of the COVID-free islands project was to offer visitors an experience resembling 'life-before-the-pandemic', by avoiding any restrictions in terms of mobility and any quarantine prior to or following their holiday (*Euronews*, 6/3/2021). The experiment entailed vaccinating the entire population of the targeted islands and limiting access to tourists from countries where a national vaccination campaign was taking place (*Athina 9.84*, 9/3/2021). The tourists were accordingly requested to present documentation certifying their relative 'immunity', in terms of either vaccination or a negative PCR test (*Le Figaro*, 26/10/2021). The international press largely welcomed the experiment, praising the Greek government for its creativity and ambition in a dramatic crisis scenario. For example, this is how The Guardian described the operation:

> Isles scattered across the Aegean archipelago are to become the first 'COVID-free' areas of Greece as vaccination efforts intensify in tourist destinations hoping for an influx of summer visitors. In one of the biggest operational challenges of modern times, authorities have vowed at least 69 islands to be fully vaccinated by the end of April.

The Guardian, 6/4/2021

The Italian press reacted even more enthusiastically. Articles in the two most widely read newspapers, *Il Corriere della Sera* and *la Repubblica* endorsed the operation, with the former publication depicting the targeted islands as a sort of new paradise in the making during the ongoing pandemic:

> Imagine a place where you can go out for a stroll with no masks, no disinfectant, and no social distancing. A place where you can kiss, dance, and go out for dinner, and perhaps meet someone by chance, without worrying about getting infected. This place would be sheltered from the effects of the global pandemic that has been paralysing all of us for over a year: a COVID-free paradise. Well, instead of just dreaming about such utopian landscapes, there is someone who is committed to making it real: the Greek Prime Minister Kyriakos Mitsotakis.

Il Corriere della Sera, 12/02/2021.

Along the same lines, *La Repubblica* ran a feature entitled *Kastellorizo, the island set free*, (13/05/2021), devoted to a Greek island where the vaccination campaign had been selectively prioritised compared to the rest of the country. The entire audio-visual report consisted of a celebration of the campaign and of the associated benefits for residents, especially in relation to economic and social life on the island.

Although no analysis of the potential impact of the experiment was available at that stage, its favourable reception was accompanied by a debate in the Italian media and some political circles concerning the potential adoption of the same approach for about thirty small islands in Italy. A formal proposal to prioritise the vaccination of island residents was therefore submitted to the Draghi government on 18 April 2021, by the mayors of the National Association of Small Island Municipalities (*Associazione Nazionale Comuni Isole Minori*; www.ancim.it), with the aim of reopening the small islands to tourists following long periods of lockdowns and limited mobility. The project was approved relatively promptly by the government, which announced the launch of a specific vaccination campaign targeting these islands beginning on 7 May 2021 (www.governo.it, 5/5/2021).

This unique campaign, driven by the desire to launch an experimental return to 'normal' tourist practices in selected enclavic environments subjected to a putative process of immunisation, clearly raised questions related to the intersections between tourism and biopolitics. Indeed, the literature on these intersections has traditionally explored the connections between tourist practices and questions of health, care, hedonism, well-being, and even 'happiness', but also bodily and social control and population management (Diken & Laustsen, 2004; Ek & Hultman, 2008; Minca, 2010; Simpson, 2016). These questions have been specifically investigated in relation to tourist enclaves and gated communities, often viewed as laboratories for social experimentation (Minca & Roelofsen, 2023). The COVID-free islands experiment thus seemed in line with this tradition, which broadly envisages the setting up of specialised tourist environments, such as tourist camps, enclavic resorts, theme parks, etc. that are regulated by specific spatio-temporal regimes and are somewhat extra-territorial in character, due to the walls that metaphorically and functionally separate them from the external environment. Such enclavic tourist spatialities have been discussed in detail by numerous tourism scholars, including Sibley (1988) and Edensor (2001), who described them as purified environments, marked not only by clear boundaries and strictly monitored access, but also by panoptic forms of surveillance and control of their internal spaces and of the subjects populating them: namely, tourists and tourism workers. Despite their secluded nature, such spaces have been analysed both in relation to the broader economic and political geographies that make them attractive and functional (Minca & Ong, 2015) and in relation to their experimental aims, including in terms of the procedures implemented to ensure the safety and control of guests and workers (Ek, 2016; Simpson, 2016).

Another key element that emerges from the literature is the insular dimension of these spaces, given that many enclavic resorts are located on islands (thus becoming 'islands-on-islands', so to speak) and given that islands are often presented as secluded spatialities—in a throw-back to the long tradition of experimental institutions (including prisons and penal colonies) based on the same principle, and equally (despite the obvious differences) aimed at the reproduction of docile subjects and bodies (Minca, 2010). Furthermore, in the Western cultural and literary tradition, islands have been often associated with the notions of 'new beginning' and 'separation' (Deleuze, 2007). In the context of promoting tourism, these notions correspond to the desire to 'get away', and radically separate oneself, from one's everyday environment, while simultaneously finding a space in which to 're-create' the individual and collective self, in line with the vast literature depicting utopian and dystopian future worlds and their societies (Minca, 2010). Accordingly, islands have long been conceptualised as an ideal spatial archetype for the realisation of fully planned tourist resorts, as conceptualised, for example, in the one-island-one-resort model promoted in the Maldives (Dell'Agnese, 2018).

It should not come as a surprise then that, in the context of a global pandemic, islands in the Mediterranean were identified as ideal environments to immunise, and as places where tourists could find not only their past freedom of movement, but also the promise of a 'new beginning', a newly conceived re-creation of the self, after a long and brutal period marked by restrictions on mobility, fear of infection and lockdowns. While attempts to implement and promote COVID-free spaces paid strong tribute to the tradition of enclavic tourist islands described above, the relationship of these spaces with the pandemic and with broader questions of immunity demands a more in-depth analysis of the biopolitical nature of these secluded tourist spaces and the related tourist experiences. We therefore now turn to the work of Italian political philosopher Roberto Esposito, who has long presented biopolitics as a manifestation of the 'immunization paradigm', which he describes as a foundational principle of governmentality in all Western societies (Esposito, 2011, 2023).

The Immunisation Paradigm and COVID-19

Esposito's analysis of the biopolitical implications of the pandemic goes beyond what he describes as "Foucault's insights" (Esposito, 2023, p. 8) into the medicalisation of society, government of the population, the expanded pastoral role of the state and the increasing adoption of dispositives of immunisation in modern Western societies. While all these biopolitical aspects were indeed implicated in the measures adopted by Western governments in response to the spread of the SARS-CoV-2 virus, Esposito has argued that, in analysing them, the specific conditions generated by an unprecedented global pandemic need to be taken into account. For this reason, he recommends examining in detail the specific ethical and biopolitical characteristics of the protocols adopted by different governments within what he defines as an overarching governmental immunisation paradigm.

According to Dempsey (2017, p. 1), Esposito's immunisation paradigm "is a biopolitical theory and a fairly recent attempt to provide a unifying vision of the biopolitics of such prophylactic procedures". Esposito understands immunisation as the enactment of a protective response to a risk that can affect a body, singular or collective, according to the dynamics of contagion. For him, through immune protection, life combats that which negates it, not via a strategy of head-on conflict but rather by means of outflanking and neutralising (Esposito, 2011, p. 8). Hence, within the immunisation paradigm, the evil that threatens the body, the object of risk, is not to be countered by keeping it away from the body, but rather by incorporating it into the body: "The dialectical figure that thus emerges is that of exclusionary inclusion or exclusion by inclusion. The body defeats a poison not by expelling it outside the organism, but by making it somehow part of the body" (ibidem). Immunity is thus defined as exemption from a risk or danger, attained via a mechanism of protection-by-negation. It is also a privilege, a unique condition that characterises an individual subject or social body. In this sense, immunity clashes with the notion of community, representing rather its opposite, given that "while community unites its members by putting them in the same condition, immunity divides its members into those who possess it and those who do not" (Esposito, 2023, pp. 185–186).

Each of the three alternative approaches to the pandemic implemented by Western governments, and identified by Esposito in his recent book *Common Immunity* (2023), bears a different relationship to the immunisation paradigm. The first approach, inspired by the notion of 'herd immunity'—and explicitly adopted in the UK, Sweden, the Netherlands, and Brazil—is described by Esposito as a form of thanatopolitics based on the marginalisation (and greater exposure to death) of the most vulnerable and 'lesser fit' segments of the population (especially the elderly) in the interests of free circulation and an active life for the most productive classes. Herd immunity, for Esposito, essentially acts similarly to autoimmune disease, because it strives to protect life by sacrificing a part of the population.

The second approach—which was implemented by most European governments, before (and to some extent after) vaccines became available—was based on so-called 'social distancing', a measure that Esposito defines as a form of negative biopolitics, because it seeks to protect from the virus by means of forced desocialisation. This form of artificial immunisation, obtained by physically separating individuals using measures of spatial segregation (lockdowns, etc.) is seen by the Italian philosopher as a paradoxical intervention, because it claims to protect the well-being of broader society by literally deconstructing its social components and their practices: "social distancing reinforces a mode of action [...] in the form of the separation of its individuals: it divides human beings in order to save a society threatened by the virus" (Esposito, 2023, p. 186).

The third approach was based on the vaccine, certainly the most powerful dispositive of immunisation, available from spring 2021 onwards. In this case, Esposito maintains, we were presented, for the first time in human history, with a potentially perfect overlap between community and immunity, given that "the vaccine, the most classic instrument of immunisation, became the most necessary common good" (Esposito, 2023, p. 187). Accordingly, for the very first time, the practices of immunisation were not conceived as "a blade that cuts through community into its selfsame form" (ibid.) but, rather, as the form that the community should adhere to:

> This turning point, inconceivable until now, was made possible not by an ethical choice [...] but by a convergence of interests that makes clear that this time, for the first time, one part of the world cannot be saved without saving all of it at the same time. Certainly, the environmental crisis has produced a similar impression in the past few decades. In this case, too, the ecological transition that it demands and makes inevitable will be either be global or not happen at all.

Esposito, 2023, p. 187.

Having said this, if one observes how the vaccine was selectively produced and distributed globally, the (affirmative) biopolitical horizon envisaged by Esposito associated with this third approach remains distant from reality. Rather, the geopolitics of the vaccine (see, among others, de Felice et al., 2023) has generated new caesurae in the global community, with significant disparities among the different countries in terms of the availability and cost of the vaccines. This represents an insurmountable obstacle to the constitution of the global immunised community theorised by Esposito. Regrettably, as matters stand: "far from coinciding with *communitas*, *immunitas* is detaching itself from it again, regressing toward its original meaning of privilege" (Esposito, 2023, p. 189). How may the COVID-free islands experiment be understood from the analytical perspective prompted by Esposito's work? Could the mass vaccination of island residents be read as an expression of affirmative biopolitics? Was the ultimate aim of producing secluded, COVID-free spaces to realise an (insular) *communitas* coinciding with *immunitas* for the sake of encouraging tourists to return? To explore some of these questions, we now turn to the case of Favignana, where the experimental immunisation of an island was implemented precisely in the name of stimulating (a largely unmonitored) tourism recovery.

Favignana: Tuna, Tuff, and Tourists

The authorities on Favignana were the first to ask for the island to be included in the experimental geography of immunisation. The Mayor of the Aegadian Islands advanced this specific request to the then president of the Sicily Region and the national government in March 2021, with the intention of implementing a strategic vaccination plan. The declared motivation behind this initiative was to relaunch the island's tourist economy in complete safety. In an interview, the mayor stated that: "Italy, like Greece, should immediately vaccinate the new COVID-free islands to make the residents immune so that [the islands], for this reason, can safely attract tourists again" (La Repubblica, 2/3/2021).

The mayor was adamant about the fact that the vaccination campaign was key to the tourist industry which:

...should recover in conditions of absolute safety. The Aegadian Islands were visited by over 60,000 tourists last year, despite the risks related to the pandemic. Once this new wave of infections is over, I anticipate a far greater increase in tourist arrivals.

Ibidem.

Foreshadowing in many ways what was to become a national debate and an associated promotion campaign, the mayor emphasised the need to quickly and comprehensively vaccinate the islands' residents to avoid placing Italy at a disadvantage to competitors such as Greece and its archipelagos. Indeed, Favignana's 4,000 residents rely on tourism (Privitera, 2020) as a key source of income and employment which, since the 1960s, has gradually been replacing the island's previous economic mainstays of tuna fishing and the extraction of tuff, a valuable material for the construction industry (Lentini, 2011, 2015). Tuna was traditionally caught using a system of fixed nets, a practice that has now been entirely abandoned, only surviving today as a form of heritage that is performed solely as a tourist attraction (Cassinelli, 2010; *Il Corriere della Sera*, 5/4/2016). Accordingly, the buildings that were once home to the tuna fish industry, namely the *Stabilimento Florio delle Tonnare di Favignana e Formica*, have been converted into an industrial heritage site that includes a museum (Inzerillo & Russo, 2013), shops, and a cocktail bar mainly frequented by tourists (*Tp24*, 16/12/2021). This site is laden with symbolic meaning for promoting the island to tourists, as insistently noted in tourism brochures and other widely circulated promotional materials including the amenity's official website: "Much of the life of the local community took place inside this vast and intricate architectural complex and hence the former Florio Plant continues to be charged with deepfelt values related to the identity of the island of Favignana and of the entire archipelago" (tonnarafloriofavignana.it, accessed 14/3/2023). The tuff quarries on the other hand have essentially been 're-naturalised' to blend into the island's broader traditional landscape and have been relabelled as *giardini ipogei* (underground gardens). The gardens are currently being promoted as a tourist attraction given their peculiar microclimate which facilitates the cultivation of fruit trees (*I Pretti Resort*, 16/5/2022). Thus, in recent decades, the seasonal presence of tourists (with about 600,000 arrivals p.a. mostly during the summer period, see *Osservatorio del turismo della Regione Siciliana*, 2022) has represented the main source of income and employment for the residents of Favignana, but has also acted as a key driver of social and cultural change on the island.

In the research presented here, we initially applied textual analysis to a range of materials such as institutional and media reports, the academic literature, documentaries, and novels. This approach was key to identifying the main narratives adopted by the authorities to attract tourists and to analysing the events that led to the implementation of a COVID-free island experiment. The first author drew on multiple ethnographic methods during her fieldwork, including semi-structured interviews, focus groups, and extensive participant observation. Three main groups of participants were identified: institutional representatives from the municipal and provincial authorities; residents and seasonal workers; and, of course, tourists. We expected that combining such a plurality of perspectives would offer a diversified image of Favignana at the time of its establishment as a COVID-free island.

As far as the participant observation was concerned, the first author approached the island context by attempting to find a seasonal job in the tourism sector on the grounds that this positioning would speed up establishing direct contact with other workers and tourists, and would also facilitate, in the long run, encounters with institutional figures. However, it soon became apparent that without local personal contacts, this was not a realistic option. Consequently, she decided to spend several weeks in Favignana during the summer of 2021, at the height of the tourist season, embodying different roles as appropriate: that of a university researcher interested in post-COVID tourism recovery; or that of a tourist/researcher comparing her own experience as a tourist with the experiences of other tourists by emulating their everyday practices.

A Fresh Start? Narratives and Tourism Practices of an Immune Paradise

We focused our investigation of Favignana as a COVID-free island on two main aspects: first, the narratives that helped to promote the island as an immune space; second, the practices of those who putatively contributed to realising this immune environment. As noted above, the literature on segregated tourist spaces has often highlighted two key notions associated with this specific geography of leisure: that of 'separation' and that of 'new beginnings'. When these notions are spatialised, they yield the utopian project of tourist enclaves. Both of these elements clearly emerged from the fieldwork conducted in Favignana, during both the participant observation and the interviews.

Narratives

The notion of 'separation', concretised in the material distance between the space of tourist islands and that of visitors' 'everyday lives', is normally associated with the creation of an atmosphere of unreality and enchantment that is well known in the realm of tourism studies. This element is a strong feature of Favignana's tourist appeal and, while it already existed prior to the pandemic, it was heavily drawn upon in the messages communicated by local authorities and businesses to promote the island as a space where visitors would be safe from the risk of contagion. The appeal to the idea that Favignana was "an outside space where one could forget one's everyday problems" emerged explicitly during the interviews with representatives of the Municipality. For instance, the local health councillor, when asked about the relationship between residents and tourists in the aftermath of COVID-19, replied:

> The residents are scared, but for the tourists, it's a bit like reality doesn't exist here. It has always been like that: not only this year with COVID, but also in the past. I think that's how it should be: people come here to stop thinking about work, about the problems they have at home, because this is what the island is like, a dimension that makes you feel good and moves your heart.

D. B., 13 August 2021.

This quote may be read as strongly evoking the notion of the island as a separate and safe place that is detached from both everyday problems and the pandemic. Indeed, this type of narrative was endorsed during several interviews with representatives of the local authorities conducted over that summer and in many other official sources.

Notably, the second element, the possibility of 'a new beginning' associated with an 'island space', was also redeployed in support of the vaccination campaign. More specifically, the island was promoted as the ideal place for, metaphorical and material, rebirth following the lockdowns, in terms of making a new start, economically, socially, and psychologically. This is why the residents' inoculations were administered on the premises of the former *Stabilimento Florio*, the industrial and archaeological tuna fishing museum (Fig. 2.1). Given the history of the former plant, which in many ways parallels the history of the island itself, it remains a key site of the social, cultural, and economic life of the island.

The repurposing of the former plant from factory to museum is aligned with a broader programme to reconvert the island's buildings for tourism activities. As borne out in the museum's illustrated brochures and during the interviews, the conversion of the former plant into a museum represented an attempt to breathe new life into a perceived bright past. The idea of rebirth is thus inherent in the museum project itself and represents an important form of cultural and economic renewal. Setting up the vaccination hub in the spaces of the former plant allowed this narrative of 'rebirth' to be further exalted and framed within the island's broader recovery in health, economic, and cultural terms.

One narrative connecting the cultural revival of the former tuna fishing and canning plant with the economic revival of the island, especially following the depression caused by the restrictions imposed on account of the epidemic, was promoted by the municipal authority, which viewed this putative connection as an incentive for both tourism and the vaccination rollout:

> We got the idea to launch this campaign at the Florio Plant because it was already a symbol of the rebirth of Favignana. The vaccine campaign also had 'rebirth' as a key word, so it was a perfect fit. I think that this

Fig. 2.1 The Vaccination Centre at the former *Stabilimento Florio* (Photograph by Caterina Ciarleglio, August 2021)

contributed to the success of the campaign; there was a very satisfactory level of participation—over 75 percent.

D. B., August 13, 2021.

Similar notions permeated the videos promoting the vaccination campaign posted on the Facebook page of the municipal authority on 11 May 2021, in which residents were urged to join the campaign in the name of a desirable and urgent 'economic, social, and psychological restart'. They were also reflected in statements by the mayor over the same period, and his invitations to residents to get vaccinated:

> ...to make our residents and islands safe, but also to offer safety to the tourists who, we hope, will wish to come and visit our beautiful island. [...] We must prepare for this new tourist season, which means work, employment, and social security for all of us.

F. F., 10 May 2021.

The idea emerging here was to leverage the supposed typical traits of the island, including the notions of 'separation' and 'new birth', to sustain a 'zero-risk' narrative that would encourage residents to live and work peacefully in contact with tourists and tourists to visit the island safely. After all, as pointed out by the then-deputy mayor:

> For us, approaching the tourist season with even a small number of people infected or with outbreaks would have meant starting off with a very serious handicap, considering that healthcare on the islands is less developed than in other locations. [...] this explains our desire, dictated by the condition of insularity, to provide safety for those who live off tourism on the islands and to give visitors the opportunity to encounter a safety framework, 'certified' by our status as a COVID-free space.

F. S., 03 August 2021.

How was this safety framework constituted and delivered? How was the COVID-free island 'certification' mentioned in this interview implemented in terms of practices? And to what extent did it influence the behaviours of tourists and residents?

Practices

During the fieldwork, it became soon apparent that, although almost all the population had been vaccinated, no other tangible factors contributed to making Favignana a COVID-free space. The attempt to build a narrative based on the canons of an island-as-a-resort may be argued to reflect a desire to emulate this kind of spatial model (Saarinen, 2019). However, our findings suggest that Favignana remained far from becoming a purified, fenced space characterised by a spatio-temporal regime imposed by an overarching apparatus of governance with the capacity to maintain hygiene, decorum, vigilance, and constant monitoring (Edensor, 2001).

The first key fact that emerged from our interviews and during our participant observation was that access to the island was not regulated by any specific measure. On the contrary, the lack of a tightly monitored external border made Favignana easily accessible throughout the tourist season by anyone who could afford to purchase a ferry ticket from the port of Trapani. The introduction of the European Digital COVID Certificate as of 6 August 2021 affected neither the possibility for tourists to travel to Favignana, nor their access to the island's restaurants, bars, or other public spaces. Unsurprisingly, this fundamental lack of control procedures was received differently by tourists and tourism workers. The generalised reaction of the former is summed up in the comment of one tourist who participated in a focus group at the beginning of August:

> Controls or no controls, I still perceive the island as COVID-free. I know the vaccine doesn't really protect me all the way, but what else do we have to do? In the end, we are on vacation.

Piazza Europa Focus Group, 7 August 2021

At the same time, the concerns expressed by one of the workers nicely reflected how the lack of measures was perceived by this group:

> We obviously aren't happy about the fact that the tourists are not subject to any form of control. The island is so tiny in any case that if there is an outbreak here everything shuts down. Even the owner [of the bar] holds you accountable: if you don't get vaccinated, you are placing the earnings of the business and the health of the people around you at risk.

Bar del Corso Focus Group, 7 August 2021

This comment should also be related to the fact that, despite the COVID-free rhetoric, the first author witnessed the almost total absence of responsible tourist behaviours such as mask wearing and social distancing, or other specific measures aimed at limiting contagion. While mandatory nationwide, these rules were weakly or patchily enforced on the island, a state of affairs that often caused unease or distress among workers forced to be in contact with tourists on a daily basis. As noted by a seasonal waiter (Fig. 2.2).

> We have not been given clear directions about how to behave.... Tourists come off the ferry in hordes, all at once, and you don't even have time to prepare and figure out how to handle them. They come in waves: one minute there is no one, the next minute the bar is full!

M., 30 July 2021

This further underscores how difficult it was to identify Favignana as a place where the authorities had imposed COVID-free measures— as one would expect after a campaign launched to attract tourists to a safe environment—or as a site with another distinctive characteristic of fully controlled tourist enclaves: that is to say, self-discipline on the part of tourists and workers (Minca, 2010). Such self-discipline, normally associated with the internalisation of specific rules or with introjected forms of panoptic control (Edensor, 2001), was essentially absent. As commented by one tourist:

> We didn't think about COVID-19 anymore... we got vaccinated and that's enough. In any case, ninety percent of the people you meet don't even wear masks...

Palazzo Florio Gardens Focus Group, 13 August 2021.

The lack of care in managing the spaces of contact among tourists and between tourists and workers was de facto perceived by many visitors not so much as a form of non-compliance, but rather as a form of reassurance. What additionally emerged during the fieldwork was that the island paradoxically lacked the high standards of hygiene and decorum that are typical of tightly controlled enclavic tourist spaces. Indeed, the local authorities were overwhelmed by the high number of tourists on the island, to the extent that even the normal maintenance of streets

Fig. 2.2 COVID-free mass tourism in Favignana (Photograph by Caterina Ciarleglio, August 2021)

and squares proved inadequate, leaving roads and beaches in a seriously neglected state—as lamented by both tourists and workers during the interviews.

In other words, although the resident population was almost entirely vaccinated, not much else was done to keep the island actually COVID-free and prevent the spread of infection. In this sense, it is difficult to argue that any "immune community", in Esposito's terms, was established on the island. The formation of such a community was paradoxically compromised by the presence of the bodies of tourists, bodies that were non-compliant given that they had not necessarily undergone vaccination, and by the absence of other "spatial tactics" (Minca, 2021) for limiting contagion. The free circulation of bodies potentially hosting the virus also denoted a relative lack of care for the health of residents, workers, and the tourists themselves—who, whether vaccinated or not, were nonetheless exposed to the risk of contracting the virus. Given that according to the *Istituto Superiore di Sanità* (the highest health authority in Italy), vaccination significantly reduced rates of hospitalisation, hospitalisation in intensive care and mortality, in relation to Esposito's proposed categories of pandemic management, Favignana may be interpreted as an example of "blunt thanatopolitics". This approach was operationalised via a form of spatial management in which the health of those who visited, worked on, and lived on the island was pre-emptively, but only weakly, protected in favour of the economic returns generated by a large influx of tourists. Thus, a form of governance was brought to bear in which the economic benefit associated with tourism was prioritised over health and safety, providing for "if not the elimination, at least the marginalisation of 'less adapted' individuals, to the benefit of the more productive segments of the population" (Esposito, 2023, p. 5).

Furthermore, our analysis pointed up a major discrepancy between actual spatial practices for pandemic containment and safety narratives. This gap suggests that the Favignana experiment exploited the narrative of an experimental immunised island-as-resort—which was key to promoting the tourist season—while essentially forgoing the distinctive control and 'purification' measures that are typical of enclavic leisure environments and that would have required a more selective approach to—and potentially limiting—tourist arrivals. Indeed, restrictions on movement or mandatory PCR testing would have potentially reduced the number of tourists and the related revenues. The relative lack of measures thus uncovers the mainly economic aims underpinning the establishment of COVID-free islands. The entire experiment might therefore be conceived more as a discursive performance or as a performative utterance (Butler, 2009) than as the realisation of an actual experimental immunised tourist space. Indeed, the recursive adoption and circulation of a narrative promoting the idea that the island was a safe place—despite the lack of effective containment measures—resulted in a set of tourist practices based on the delusionary belief that Favignana was indeed the promised immune paradise. It was 'as if' crossing the natural boundaries of the island allowed the bodies of tourists, residents, and workers to automatically benefit from a condition of immunity, guaranteed by the normative discourse endorsed by the local authorities and the tourist industry. 'As if', on entering the supposedly immune space of the island—a place that discursively linked immunity with the temporary state of being 'happy people'—necessarily entailed the metamorphosis of tourists into 'healthy people'.

The case study of COVID-free Favignana thus reveals how the tourism industry was reorganised very rapidly to subsume even the pandemic emergency according to the logic of capitalist accumulation in the form of leisure consumption. In many ways, the COVID-free islands project was an integral part of this reorganisation. Indeed, it is well known that the tourist industry "commodifies all obstacles in its path" (Oakes & Minca, 2004, p. 288) by readjusting and co-opting sociopolitical contradictions so that they can be reincorporated into a system of value production. Accordingly, the case study discussed here may be read as an experiment in the economic recovery of a space where tourism is the primary source of livelihood. It was a biopolitical experiment that, by adopting a visionary narrative associated with the setting up of utopian, COVID-free tourist spaces, exposed tourists, but also residents and workers, to a potentially higher risk of becoming infected by the virus.

Conclusions

The findings of our investigation into the Favignana COVID-free experiment may be read in light of the key starting assumptions of our research project, and specifically, in light of Esposito's caution that, in order to implement affirmative biopolitics in response to the unprecedented threat of a global pandemic, all selective biopolitical 'cuts' effected on the population to further particular economic interests should be abandoned in favour of strategies aimed at achieving a difficult but highly desirable coincidence between a *communitas* and an *immunitas*. The primacy of tourism recovery, far from facilitating the attainment of an immune community—which would have represented an affirmative form of biopolitics—somehow turned the Favignana experiment upon its head, transforming the island into a space regulated by a bland form of thanatopolitics, while it nevertheless continued to be promoted as 'immune' via narratives aimed at drawing the tourists back.

However, it should also be emphasised that the very concept of a segregated island-as-a-resort is constitutively at odds with the coexistence of community and immunity as conceptualised by Esposito. The spatialities of enclavic resorts are such that—by definition—they receive the bodies of highly selected tourists, bodies whose very presence tends to be an expression of privilege. Segregated spaces of tourism, as forms of spatial confinement, constitutively operate a 'cut', a caesura, in the (global) community. More specifically, they are constructed along a line that separates the 'frontstage' (for selected paying tourists) from other tourists, residents, and the 'backstage' (the resort's workers) (Edensor, 2001). In the case of the COVID-free islands, the separation between residents and tourists was accomplished by means of vaccination, thus rendering the immunisation paradigm as a manifestation of yet another privilege. Read through the biopolitical lens offered by Esposito, the Favignana case study represents but another missed opportunity to rethink immunity as a universal and highly desirable objective. It demonstrates yet again that as long as enclavic tourist spaces continue to be promoted and managed as sites of privileged consumption, they will reproduce the selective logic of a biopolitical caesura that cuts into the body of the broader population; a caesura that in the case of Favignana, took the form of reproducing unequal levels of exposure and protection between tourists, residents, and tourism workers.

Competing Interests

The authors have no conflicts of interest to declare that are relevant to the content of this chapter.

References

Aime, M., & Papotti, D. (2012). *L'altro e l'altrove. Antropologia, geografia e turismo.* Einaudi.

Ajana, B. (2021). Immunitarianism: Defence and sacrifice in the politics of COVID-19. *History and Philosophy of the Life Sciences, 43*(25), 1–31. https://doi.org/10.1007/s40656-021-00384-9

Aronica, M., Pizzuto, P., & Sciortino, C. (2021). COVID-19 and tourism: What can we learn from the past? *World Economy, 45*(2), 430–444. https://doi.org/10.111/twec.13157

Barker, J., & Rodway-Dyer, S. (2022). The elephant in the zoom: The role of virtual safaris during the COVID-19 pandemic for conservation resilience. *Current Issues in Tourism, 26*(13), 2221–2234. https://doi.org/10.1080/13683500.2022.2132921

Bichler, B., Petry, T., & Peters, M. (2021). 'We did everything we could': How employees' made sense of COVID-19 in the tourism and hospitality industry. *Current Issues in Tourism, 25*(23), 3766–3782. https://doi.org/10.1080/13683500.2021.1985974

Brune, S., Knollenberg, W., & Vilá, O. (2023). Agritourism resilience during the COVID-19 crisis. *Annals of Tourism Research, 99*, Article 103538. https://doi.org/10.1016/j.annals.2023.103538

Butcher, J. (2021). Debating tourism degrowth post COVID-19. *Annals of Tourism Research, 89*, Article 103205. https://doi.org/10.1016/j.annals.2021.103250

Butler, J. (2009). *Ces corps qui comptent: de la matérialité et des limites discursives du sexe.* Editions Amsterdam.

Cassinelli, E. (2010). *Tonni turisti e tonnaroti. L'isola mediterranea fra immaginario turistico e strategie locali: il caso di Favignana.* Tesi di dottorato, Università Milano-Bicocca.

Castañeda-García, J. A., Sabiote-Ortiz, C. M., Vena-Oya, J., & Epstein, D. M. (2022). Meeting public health objectives and supporting the resumption of tourist activity through COVID-19: A triangular perspective. *Current Issues in Tourism, 26*(10), 1617–1634. https://doi.org/10.1080/13683500.2022.2062306

Chan, Y. W., & Haines, D. (2022). Diseasescape and immobility governance: COVID-19 and its aftermaths. *Mobilities, 1–16.* https://doi.org/10.1080/17450101.2022.2150560

Cheng, L., & Liu, L. (2022). Exploring posttraumatic growth after the COVID-19 pandemic. *Tourism Management, 90*(1–12), 104474. https://doi.org/10.1016/j.tourman.2021.104474

Connell, J. (2021). COVID-19 and tourism in Pacific SIDS: Lessons from Fiji, Vanuatu and Samoa? *The round Table, 110*(1), 149–158. https://doi.org/10.1080/00358533.2021.1875721

Collins-Kreiner, N., & Ram, Y. (2020). National tourism strategies during the Covid-19 pandemic. *Annals of Tourism Research, 89*, 1–6. https://doi.org/10.1016/j.annals.2020.103076

Cresswell, T. (2020). Valuing mobility in a post COVID-19 world. *Mobilities, 16*(1), 51–65. https://doi.org/10.1080/17450101.2020.1863550

de Felice, G., Tutal, N., & Sciaraffa, N. (2023). Geopolitical aspects of COVID-19 vaccines distribution. *European Journal of Environment and Health, 7*(3), Article em0132. https://doi.org/10.29333/ejeph/12779

Deleuze, G. (1951). *L'isola deserta e altri scritti.* Einaudi.

Deleuze, G. (2007). *Cause e ragioni delle isole deserte,* In D. Borca (Ed.), *L'isola deserta e altri scritti. Testi e interviste 1953–1974* (pp. 3–9). Einaudi.

Dempsey, O. (2017). Screening for cancer risk: Esposito's immunization paradigm, capitalism and the logic of fantasy. *Annual Review of Critical Psychology, 13.*

dell'Agnese, E. (2018). One island, one resort. Il turismo enclave alle Maldive come eterotopia pianificata. *Bollettino della Società Geografica Italiana, 14*(1), 27–39. https://doi.org/10.13128/bsgi.v1i1

de Spuches, G., Sabatini, F., Palermo, G., & Caravello, E. (2020). Risk narrations and perceptions in the COVID-19 time. A discourse analysis through the Italian press. *Aims Geosciences*, 6(4), 504–514. https://doi.org/10.3934/geo sci.2020028

Diken, B., & Laustsen, C. B. (2004). Sea, sun, sex and the discontents of pleasure. *Tourist Studies*, 4(2), 99–114. https://doi.org/10.1177/146879760 4054376

Dybsand, H. N. H. (2022). 'The next best thing to being there'–participant perceptions of virtual guided tours offered during the COVID-19 pandemic. *Current Issues in Tourism*, 1–14. https://doi.org/10.1080/13683500.2022. 2122417

Edensor, T. (2001). Performing tourism, staging tourism: (Re)producing tourist space and practice. *Tourist Studies*, 1, 59–82. https://doi.org/10.1177/146 879760100100104

Ek, R. (2016). The tourist camp: All-inclusive tourism, hedonism and biopolitics. In *Privileged mobilities: Tourism as world ordering* (pp. 47–66). Cambridge Scholars Publishing.

Ek, R., & Hultman, J. (2008). Sticky landscapes and smooth experiences: The biopower of tourism mobilities in the Öresund region. *Mobilities*, 3(2), 223–242. https://doi.org/10.1080/17450100802095312

Ek, R., & Tesfahuney, M. (2019). Topologies of tourism enclaves. *Tourism Geographies*, 21(5), 864–880. https://doi.org/10.1080/14616688.2019. 1663910

Esposito, R. (2011). *Immunitas*. Polity Press.

Esposito, R. (2023). *Common immunity. Biopolitics in the age of the pandemic*. Polity Press.

Esposito, R., & Campbell, T. (2006). The immunization paradigm. *Diacritics*, 36(2), 23–48. https://doi.org/10.1353/dia.2008.0015

García, D., & Matías M. (2022). Domestic tourism and the resilience of hotel demand. *Annals of Tourism Research*, 93, Article 103352. https://doi.org/ 10.1016/j.annals.2022.103352

Gössling, S., & Schweiggart, N. (2022). Two years of COVID-19 and tourism: What we learned, and what we should have learned. *Journal of Sustainable Tourism*, 30(4), 915–931. https://doi.org/10.1080/09669582.2022. 2029872

Gu, Y., Bhakti, S., Onggo, M., Kunc, H., & Bayer, S. (2021). Small Island Developing States (SIDS) COVID-19 post-pandemic tourism recovery: A system dynamics approach. *Current Issues in Tourism*, 25(9), 1481–1508. https:// doi.org/10.1080/13683500.2021.1924636

Hall, C. M., Scott, D., & Gössling, S. (2020). Pandemics, transformations and tourism: Be careful what you wish for. *Tourism Geographies*, 22(3), 577–598. https://doi.org/10.1080/14616688.2020.1759131

Haywood, K. M. (2020). A post COVID-19 future-tourism re-imagined and re-enabled. *Tourism Geographies, 22*(3), 599–609. https://doi.org/10.1080/14616688.2020.1762120

Higgins-Desbiolles, F. (2020). The 'war over tourism': Challenges to sustainable tourism in the tourism academy after COVID-19. *Journal of Sustainable Tourism, 29*(4), 551–569. https://doi.org/10.1080/09669582.2020.1803334

Hopkins, D. (2021). Crises and tourism mobilities. *Journal of Sustainable Tourism, 29*(9), 1423–1435. https://doi.org/10.1080/09669582.2021.1905969

Hüsser, A. P., Ohnmacht, T., & Thao, V. T. (2023). Tourists' preventive travel behaviour during COVID-19: The mediating role of attitudes towards the intention to apply non-pharmaceutical interventions (NPIs) while travelling. *Current Issues in Tourism, 1–15*. https://doi.org/10.1080/13683500.2022.2162373

Inzerillo, B., & Russo, D. (2013). Il progetto dell'immateriale per la valorizzazione delle identità culturali. L'ex Stabilimento Florio di Favignana. In *Il design dei beni culturali, crisi, territorio, identità* (pp. 162–173). Rizzoli.

Ioannides, D., & Gyimóthy, S. (2020). The COVID-19 crisis as an opportunity for escaping the unsustainable global tourism path. *Tourism Geographies, 22*(3), 624–632. https://doi.org/10.1080/14616688.2020.1763445

Jensen, O. B. (2021). Pandemic disruption, extended bodies, and elastic situations—Reflections on COVID-19 and mobilities. *Mobilities, 16*(1), 66–80. https://doi.org/10.1080/17450101.2021.1867296

Khalid, U., Emeka Okafor, L., & Burzynska, K. (2021). Does the size of the tourism sector influence the economic policy response to the COVID-19 pandemic? *Current Issues in Tourism, 24*(19), 2801–2820. https://doi.org/10.1080/13683500.2021.1874311

Kowalewski, M. (2021). Walking Rome without leaving home: Practicing cultural geography during the COVID-19 pandemic. *Cultural Geographies 28*(3), 563–567. https://doi.org/10.1177/1474474021993417

Lentini, R. (2011). *Il cantu dei pirriaturi.* In M. Modica (Ed.), *Favignana tra mare e terra* (pp. 9–14). Fondazione Ignazio Puttitta.

Lentini, R. (2015). *La rivoluzione di latta. Breve storia della pesca e dell'industria del tonno nella Favignana dei Flor.* Edizioni Torre del Vento.

Lew, A., Cheer, J. M., Haywood, M., Brouder, P., & Salazar, N. B. (2021). Visions of travel and tourism after the global COVID-19 transformation of 2020. *Tourism Geographies, 22*(3), 455–466. https://doi.org/10.1080/14616688.2020.1770326

Lin, W., Yeoh, S. A., & B. (2021). Pathological (Im)mobilities: Managing risk in a time of pandemics. *Mobilities, 16*(1), 96–112. https://doi.org/10.1080/17450101.2020.1862454

Lisle, D. (2016). Exotic endurance: Tourism, fitness and the Marathon des Sables. *Environment and Planning d: Society and Space, 34*(2), 263–281. https://doi.org/10.1177/0263775815613094

Minca, C. (2006). Reinventing the square. In C. Minca & T. Oakes (Eds.), *Travels in paradox: Remapping tourism* (pp. 145–174). Rowman & Littlefield.

Minca, C. (2010). The island: Work, tourism and the biopolitical. *Tourist Studies, 9*(2), 88–108. https://doi.org/10.1177/1468797609360599

Minca, C. (2011). No country for old men. In C. Minca & T. Oakes (Eds.), *Real tourism: Practice, care and politics in contemporary travel culture* (pp. 12–37). Routledge.

Minca, C. (2021). Tattiche spaziali e emergenza: Qualche riflessione su biopolitica, mobilità e soggetto. In G. Iacoli, D. Papotti, G. Peterle, & L. Quaquarelli (Eds.), *Culture della mobilità: Immaginazioni, rotture, riappropriazioni del movimento* (pp. 159–171). Franco Cesati Editore.

Minca, C., & Ong, C. E. (2015). Hotel California: Biopowering tourism, from New Economy Singapore to Post-Mao China. In G. T. Jóhannesson, C. Ren, & R. van der Duim (Eds.), *Tourism encounters and controversies: Ontological politics of tourism development* (pp. 159–180). (New directions in tourism analysis). Ashgate.

Minca, C., & Roelofsen, M. (2019). Becoming airbnbeings: On datafication and the quantified self in tourism. *Tourism Geographies, 23*(4), 743–764. https://doi.org/10.1080/14616688.2019.1686767

Minca, C., & Roelofsen, M. (2023). Post-COVID biopolitical fantasies and the case of the Dutch 'Pilot Holidays', *Environment and Planning C: Politics and Space*, online first.

Mostafanezhad, M. (2020). Covid-19 is an unnatural disaster: Hope in revelatory moments of crisis. *Tourism Geographies, 22*(3), 639–645. https://doi.org/10.1080/14616688.2020.1763446

Nikolaeva, A. (2023). Living without commuting: Experiences of a less mobile life under COVID-19. *Mobilities, 18*(1), 1–20. https://doi.org/10.1080/17450101.2022.2072231

Nuno, A., & Paulo, R. (2020). March 2020: 31 days that will reshape tourism. *Current Issues in Tourism, 24*(19), 2768–2783. https://doi.org/10.1080/13683500.2020.1863927

Oakes, T. (2020). Afterword: A critical reckoning with the 'Asian Century' in the shadow of the Anthropocene. *Tourism Geographies, 23*(4), 937–943. https://doi.org/10.1080/14616688.2020.1833973

Oakes, T., & Minca, C. (2004). Tourism, modernity, and postmodernity. In A. Lew, M. Hall & A. M. (Eds.), *The Wiley Blackwell Companion to tourism* (pp. 281–289). https://doi.org/10.1002/9780470752272.ch22

Paddison, B., & Hall, J. (2022). Tourism policy, spatial justice and COVID-19: Lessons from a tourist-historic city. *Journal of Sustainable Tourism, 1–16.* https://doi.org/10.1080/09669582.2022.2095391

Parkers, W. F., Bay, M., Bryce, I. (Producers), & Bay, M. (Director). (2005). *The Island* [Film]. USA: Dreamworks Pictures, 2005.

Privitera, D. (2020). Turismo lento e territori insulari. Il caso studio Favignana. *Bollettino dell'Associazione Italiana di Cartografia, 169,* 145–153.

Roelofsen, M., & Minca, C. (2018). The superhost. *Geoforum, 91,* 170–181. https://doi.org/10.1016/j.geoforum.2018.02.021

Saarinen, J., & Wall-Reinius, S. (2019). Enclaves in tourism: Producing and governing exclusive spaces for tourism. *Tourism Geographies, 21*(5), 739–748. https://doi.org/10.1080/14616688.2019.1668051

Salazar, N. B. (2021). Existential vs. essential mobilities: Insights from before, during and after a crisis, *Mobilities, 16*(1), 20–34. https://doi.org/10.1080/17450101.2020.1866320

Samdin, Z., Abdullah, S. I. N. W., Khaw, A., & Subramaniam, T. (2022). Travel risk in the ecotourism industry amid COVID-19 pandemic: Ecotourists' perceptions. *Journal of Ecotourism, 21*(3), 266–294. https://doi.org/10.1080/14724049.2021.1938089

Seyfi, S., Hall, C. M., & Shabani, B. (2020). COVID-19 and international travel restrictions: The geopolitics of health and tourism. *Tourism Geographies, 25*(1), 357–373. https://doi.org/10.1080/14724049.2021.1938089

Sibley, D. (1988). Survey 13: Purification of space. *Environment and Planning D: Society and Space, 6*(4), 409–421. https://doi.org/10.1068/d060409

Simpson, T. (2016). Tourist utopias: Biopolitics and the genealogy of the post-world tourist city. *Current Issues in Tourism, 19*(1), 27–59. https://doi.org/10.1080/13683500.2015.1005579

Tan, D., & Caponecchia, C. (2021). COVID-19 and the public perception of travel insurance. *Annals of Tourism Research, 90,* Article 103106. https://doi.org/10.1016/j.annals.2020.103106

Turri, E. (1998). *Il paesaggio come teatro.* Marsilio.

Veijola, S., & Jokinen, E. (1994). The body in tourism. *Theory Culture & Society, 11*(3), 125–151. https://doi.org/10.1177/026327694011003006

Williams, C. (2021). Impacts of the coronavirus pandemic on Europe's tourism industry: Addressing tourism enterprises and workers in the undeclared economy. *International Journal of Tourism Research, 23*(1), 79–88. https://doi.org/10.1002/jtr.2395

Xiang, K., Huang, W. J., Gao, F., & Lai, Q. (2022). COVID-19 prevention in hotels: Ritualized host-guest interactions. *Annals of Tourism Research, 93,* Article 103376. https://doi.org/10.1016/j.annals.2022.103376

Zhu, O. Y., Grün, B., & Dolnicar S. (2022). Tourism and vaccine hesitancy. *Annals of Tourism Research, 92*, Article 103320. https://doi.org/10.1016/j.annals.2021.103320

SITOGRAPHY

Associazione Nazionale Comuni Isole Minori. (2021, April 14). *Comunicato stampa ANCIM situazione vaccini.* Retrieved on August 16, 2023], from https://www.ancim.it/comunicato-ancim-situazione-vaccini.htm

Athina 9.84. (2021, February 28). *Το Καστελλόριζο είναι Covid-free – Όλοι οι κάτοικοι εμβολιάσθηκαν.* Retrieved on August 21, 2023, from https://athina984.gr/2021/02/28/corriere-della-sera-to-kastellorizo-einai-covid-free-oloi-oi-katoikoi-emvoliasthikan/

Casadio, G. (2021, March 3). *Il sindaco di Favignana scrive a Figliuolo.* La Repubblica. Retrieved on August 21, 2023, from https://www.repubblica.it/politica/2021/03/02/news/vaccini_sindaco_favignana_piccole_isole_covid_free289854919/?callback=in&code=YWRMMMZMNTYTMZGWMI0ZO TNJLTLHMWYTODI5YWZKY2I5M2Q2&state=7ec35aac3220458382db9 2ec6653cd20

Dartford, K., & Staikos, A. (2021, March 6). *This little-known Greek island has become Europe's first COVID-free destination.* Euronews. Retrieved on August 21, 2023, from https://www.euronews.com/travel/2021/03/05/this-little-known-greek-island-has-become-europe-s-first-covid-free-destination

Governo Italiano. Presidenza del Consiglio dei Ministri. (2021, May 5). *Avviato il piano esistente per la vaccinazione delle isole minori.* Presidenza del Consiglio dei ministri. Retrieved on August 16, 2023, from https://www.governo.it/it/dipartimenti/commissario-straordinario-lemergenza-covid-19/16786

I Pretti Resort. (2020). *Le cave di tufo: i "pirrituri" favignanesi.* Retrieved on August 16, 2023, from https://www.iprettiresort.it/favignana/cosa-visitare-a-favignana/le-cave-di-tufo-e-i-giardini-ipogei-di-favignana/

Kaniadakis, E. (2021, May 14). *Greece is preparing for tourist influx—but is it ready?* Euronews. Retrieved on August 21, 2023, from https://www.euronews.com/2021/05/13/greece-is-preparing-for-a-tourist-influx-but-is-it-ready

Kefalas, A. (2021, October 26). *Voyage en Grèce et Covid-19: formulaire, passe sanitaire, tests PCR.* Le Figaro. Retrieved on August 21, 2023, from https://www.lefigaro.fr/voyages/voyage-en-grece-covid-passe-sanitaire-formulaire-20211026

Libero consorzio comunale di Trapani. (2021). *Turismo in cifre.* Retrieved on August 16, 2023, from http://www.consorziocomunale.trapani.it/provinciatp/zf/index.php/servizi-aggiuntivi/index/index/idtesto/230

Nicastro, A. (2021, February 12). *Il piano della Grecia per diventare Covid free (e le isole che lo sono già)*. Il Corriere. Retrieved on August 21, 2021, from https://www.corriere.it/esteri/21_febbraio_12/piano-grecia-divent are-covid-free-isole-che-sono-gia-dbddba4a-6d19-11eb-9243-a33dd4e4072e. shtml

Segond, V., Fayard, Q., & Kefalas, A. (2021, May 11). *Ces îles «sans Covid» qui veulent rassurer les touristes*. Le Figaro. Retrieved on August 21, 2023, from https://www.lefigaro.fr/international/ces-iles-sans-covid-qui-veu lent-rassurer-les-touristes-20210511

Smith, H. (2021, April 6). *Aegean islands aim to become first fully vaccinated areas of Greece*. The Guardian. Retrieved on August 21, 2023, from https://www.theguardian.com/world/2021/apr/06/aegean-isl ands-to-become-first-covid-free-areas-of-greece

Tonnara Florio di Favignana. (2020). *Rete di spazi e paesaggi delle isole Egadi*. Retrieved on August 16, 2023, from https://www.tonnarafloriofavignana.it/ ex-stabilimento/

Tp24. (2021, December 16). *Il manager di Singapore con interessi a Favignana, perquisizioni e sequestri della Finanza*. Retrieved on August 21, 2023, from https://www.tp24.it/2021/12/16/cronaca/il-manager-di-singapore-con-interessi-a-favignana-perquisizioni-e-sequestri-della-finanza/171934

Walsh, D. (2021, April 14). *These small Italian islands could be COVID-free in time for summer holidays*. Euronews. Retrieved on August 21, 2023, from https://www.euronews.com/travel/2021/04/14/these-small-italian-isl ands-could-be-covid-free-in-time-for-summer-holidays

CHAPTER 3

Between Threat and Privilege: Narratives of Tourism in Crisis

Myra Coulter and *Dominic Lapointe*

Abstract In this chapter we delve into the role of tourist mobilities in the propagation of the coronavirus, and the related undoing of dominant representations of exclusive leisure spaces and privileged mobilities. Specifically, we examine the collapse of the international travel and tourism industry and the reordering of uneven (im)mobilities in the weeks leading up to March 11, 2020. We draw upon the discourses and experiences of a range of stakeholders, or para-informants, in the tourism value chain, as represented in online news media accounts. Through a hybrid discourse and narrative analysis of digital news articles, our chapter highlights the biopolitical dimensions of tourism that became exposed through the COVID-19 pandemic, focusing on the tourist subject, and states and spaces of exception at multiple scales. Our analysis of the

M. Coulter · D. Lapointe (✉)
Department of Urban and Tourism Studies, Université du Québec à Montréal, Montreal, QC, Canada
e-mail: lapointe.dominic@uqam.ca

M. Coulter
e-mail: o_neill_coulter.myra_jane@courrier.uqam.ca

M. Roelofsen and C. Minca (eds.), *Tourism and Biopolitics in Pandemic Times*, https://doi.org/10.1007/978-3-031-46399-0_3

pandemic's becoming phase in early 2020 points up multiple – social, spatial, and political – emergent tendencies in the global tourism system.

Keywords Para-informant · Tourist-subject · State of exception · Privileged mobilities

THE BECOMING OF THE PANDEMIC

As soon as the World Health Organization (WHO) declared that the COVID-19 outbreak had attained pandemic status, prompting the enforcement of lockdown and mobility restrictions around the world, tourism studies scholars began writing about tourism and the pandemic, largely focusing on the tourist sector's potential transformation or its eventual recovery (Brouder et al., 2020; Kadri et al., 2023). This was followed by a plethora of scientific papers addressing policy responses, the strategies deployed by businesses, and the impact of lockdowns on tourism. While this literature is of interest, it tends to overlook the initial weeks when the pandemic was just coming into being: specifically, those immediately prior to the announcement of the pandemic on 11 March 2020 and the consequent brutal halting of leisure mobilities in most countries. This is the period that we set out to illuminate in this chapter. We delve into the role of tourist mobilities in the propagation of the coronavirus, and the related undoing of dominant representations of exclusive leisure spaces and privileged mobilities. More precisely, we examine the collapse of the international travel and tourism industry and the reordering of uneven (im)mobilities in the weeks leading up to 11 March 2020. To this end, we draw upon the discourses and experiences of a range of stakeholders in the tourism value chain, as represented in online news media accounts. Our aim in this analysis is to unpack the biopolitical dimension of tourism as revealed through the becoming of the pandemic.

Biopolitics is the intersection between life and politics (Campbell & Sitze, 2013) and wields the power to manage individuals and populations (Foucault, 1997, 2004). In the context of tourism, biopolitics involves the ordering and control of (im)mobility (Ek & Hultman, 2008; Sheller, 2016), and the various enclavic spaces deployed to enable and regulate tourism (Ek & Tesfahuney, 2019; Minca, 2010; 2015). Thus, the biopolitical management of tourism echoes Agamben's (1998) state of exception, as well as the interaction between bare life and sociopolitical life. Furthermore, biopolitical understandings of tourism envisage both immunization strategies (Esposito, 2011) for protecting tourism from the outside world by means of leisure enclaves and attempts at self-immunization in terms of offsetting the impacts of tourism through even more tourism (Sarrasin & Lapointe, 2021). As we shall see, these dimensions of biopolitics were also expressed in the sanitary measures introduced in early 2020 in an effort to stem the becoming of the pandemic, in the characterization of bodies in relation to risk, in the 'cuts' inflicted upon the social body (and the touristic body), and in the various states of exception.

The becoming of the pandemic, and the shock that this imparted to tourism mobilities, also reflected the outlook of neoliberal subjects and their internalization of the market as their rule of conduct (Brown, 2015). Following Foucault (2007) in *Securité, Territoire, Population*, the management of the population in the build-up to the pandemic may be seen as informed by the two-way relationship between population and production, whereby contemporary Western states control their populations because people are a factor in productivity. Indeed, this power over populations reflects what Foucault (2007) termed the logic of circulation: the circulation of capital, man(*sic*)power, goods, etc. In the context of tourism, the biopolitics of circulation or of what we nowadays call mobility, plays an even greater part, given that the circulation of bodies simultaneously represents both production *and* market, merged into what has been referred to as the touristic body (Lapointe, 2022). In early 2020, the emergence of a direct relationship between the voluntary circulation of bodies and the spread of the virus responsible for COVID-19 led to a fundamental clash between the maintenance of leisure mobilities and the protection of populations, that is to say, between the preservation of the tourism economy and the safety of the public. In the context of rapidly unfolding practices, policies, and strategies for safeguarding both the travel and tourism industry and public health, the mass media became an invaluable source[1] for investigating the biopolitical dimensions

---

[1] The terms 'misinformation' and 'disinformation' became associated with the abundance of information produced during the early phases of the COVID-19 pandemic. We are not interested here in the veracity of media accounts but, rather, in their representations of tourism and travel during that specific period.

of tourism. In this chapter, we draw on a corpus of online media articles that represented the pandemic's coming into being to examine the discourses and experiences of stakeholders around the world in relation to a series of 'newsworthy' events.

A Hybrid Discourse and Narrative Analytical Approach

Ordinary tourism activities are typically not newsworthy. Indeed, mass tourism and travel writing is often relegated to specialized columns and reviews, whose aim is to entice prospective consumers into imagining and exploring a range of leisure experiences. However, the period directly preceding the declaration of the COVID-19 pandemic provided an opportunity for a wide range of tourism value chain stakeholders to be featured in the generalist media. Thus, following the implementation of strict public health measures in Canada, where both authors of this chapter live and work, we gathered textual data from multiple sources including media websites and both the Factiva and ProQuest databases. We compiled digital news articles published by four international, English-language media outlets between 1 February and 31 March 2020, attaining a corpus of 1818 articles. We chose the media outlets (1) South China Morning Post (SCMP), Hong Kong, (2) Asahi Shimbun, Japan, (3) New York Times (NYT), USA, and (4) Globe and Mail (G&M), Canada, given that these are all local daily newspapers with a global reach, and with a view to tracking the evolution of the public health threat as it moved from East Asia towards Europe and North America. We selected articles based on simultaneous keyword and thematic searches, retaining texts that contained at least one match for each category of search term. The keywords included '(corona)virus', 'covid', 'outbreak', and 'pandemic', while the themes centred around tourism, travel, mobility, and hospitality. We subsequently manually reviewed the corpus for salience, eliminating duplicates and purely descriptive texts (e.g., COVID-19 case counts).

We adopted a hybrid discourse and narrative analytical approach to examine the accounts of a range of key tourism stakeholders, including travellers, authority figures, tourism industry representatives, tourism sector workers, and, indeed, journalists. In this chapter, we home in on four areas of interest, namely: (1) 'experiences' of spatial and social hierarchies, including uneven mobilities; (2) the 'voices' that were featured and those that were silenced; (3) strategies used to manage the public health threat and the threat to tourism; and (4) novel tourism discourses and practices. In many instances, the reporters had engaged directly with informants via telephone interviews, social media, or email exchanges, but some of the narratives were second-hand, having been earlier reported elsewhere or posted on social media. In general, we view these first-person accounts as ancillary to the direct participation that is typical of ethnographic research, but nevertheless as revelatory of lived experiences during the becoming of the pandemic. Therefore, we took the tourism stakeholders engaged with by reporters to be para-informants and analysed their narratives with a view to advancing understanding of the biopolitics of tourism. We complemented our analysis of the narrative corpus by also considering the broader cultural, political, and economic conditions at the time the pandemic was coming into being, focusing on the events that marked the period from 1 February 2020 until the announcement of the pandemic around six weeks later. In the following sections, we present our findings in the form of five narrative discourses that exemplify key biopolitical phenomena during our timeframe of interest.

FLOATING PRISON OR... LOVELY EXPERIENCE?

The spatial arrangement of amenities and passengers on a cruise ship is conducive to the spread of viral and bacterial infections such as rotavirus, norovirus, or E. coli (Arellano & Nasab, 2020; Renaud, 2020). During the first two weeks of February 2020, numerous reports were published of passengers stranded on vessels in the Pacific region, due to efforts to contain the spread of the novel coronavirus following its detection onboard. Various cruise ships, most famously the Diamond Princess, were likened to a 'floating prison' or a 'petri dish', in allusion to the risk of exposure to the coronavirus and mistrust of the authorities who were managing the disaster. Nevertheless, the degree of health risk to different passengers and, indeed, crew members was highly uneven, and the social and spatial configurations that accounted for this hierarchy are reflected in the media accounts. To this effect, British passenger, David Abel was reported in the SCMP as feeling "really sorry [for those] with inside cabins who've got no natural light, no fresh air. It's going to be pretty grim for them for two weeks" (Associated Press, 2020, Feb. 7). In contrast, some passengers were able to enjoy the experience.

Among the more privileged quarantined passengers, Zahra Jennings, a retired staff nurse from Britain, remarked "Everyone says 'poor you' […] But there was no poor you. We had free internet and free wine. We had three-course meals. There was so much choice […] The experience was 'lovely'" (Reuters & DPA, 2020, Feb. 15). However, the spatial disposition of the cruise ship passengers and, by extension, of the cruise ship itself is a factor in the relative privilege of those onboard. Highlighting the impermanence of the spatial ordering, passenger Mr. Fehrenbacher was quoted in the G&M as noting that "the ship has docked differently […] Our view changed from the city of Yokohama [to] a slew of ambulances, fire trucks, reporters, pallets [and] a bunch of people in [hazardous materials] suits. It's a little bit of a different vibe on this end" (Globe & Mail, 2020, Feb. 13). Therefore, despite the passengers' subjugation to days or weeks of confinement, and the uncertainty or boredom experienced by many of them, the glimpses provided by reporters into their daily lives reflect a generalized ambivalence towards the health threat. Indeed, some experienced the cruise experience as normal, albeit a bit longer than anticipated. For one intrepid passenger, retired Canadian aerospace engineer Pierre Ashby, the absence of structure and entertainment provided "[his] best cruise ever […] 'Usually you buy a cruise, and you know exactly what you are going to get […] This was an adventure'" (Reuters & DPA, 2020, Feb. 15).

Limited attention was paid to the experience and testimonies of the hundreds or thousands of crew members who were onboard these vessels alongside the passengers. Like other branches of the tourism industry, the cruise sector is highly labour intensive, with large numbers of frontline workers whose direct contact with passengers puts them at increased risk of contracting disease (Rose, 2021). One unnamed woman working in the kitchens who tested positive for the virus is documented as saying, "[t]he emotional, psychological and physical stress that we are going through now is really hard" (Rich, 2020, Feb. 22). Overall, however, the accounts of crew members threatened by COVID-19 and contained on cruise ships for extended periods of time are virtually absent from our corpus. In contrast, the misadventures of infected passengers—including John Haering, 63, a retired operations manager with the Union Pacific Railroad and resident in Tooele, Utah, USA (Rich & Wong, 2020, Feb. 17)—received greater attention. In this regard, the contaminated and quarantined cruise ship acted as a microcosm of broader society, where social and spatial hierarchies would prove to influence lived experiences of the public health crisis on the part of unevenly positioned subjects (Arellano & Nasab, 2020).

THE TOURIST-SUBJECT

The significance of home as an integral component of privileged mobile subjectivities is highlighted throughout our corpus, represented as both the place from which travel begins and the place to which travellers return. Indeed, in the absence of a homeplace, travellers lose their status as tourists, becoming migrants or vagabonds (Bauman, 1996). By extension, and in other words, home is the space between trips. To this effect, NYT contributing author Reif Larsen observed that, "[...] we travel to force ourselves to take a breath, to bend space and time, even if just for a moment. We go there so we can come back and appreciate the here" (Larsen, 2020, Mar. 24). Thus, the cultivation of a positive relationship with home is the very reason for travel. This privileged relationship with home, however, is uneven and, in the context of evacuation and repatriation strategies for reducing the risk of undergoing lockdown or of contracting disease, the media accounts emphasized the privilege of holding the 'right' citizenship as a condition for a privileged relationship with home.

Sarah Arana, 52, a medical social worker from Paso Robles, California, was blown away by the "phenomenal amount of resources" mobilized to receive evacuees, like her, back home. According to NYT reporters Rich and Wong (2020, Feb. 17), "[w]hen one of the planes landed in California, a line of officials from the military, the Centers for Disease Control and Prevention, and the Department of Homeland Security greeted passengers with banners that read 'Welcome home'". However, the repatriation efforts of different nation-states were perceived to be uneven, with North Americans reporting feeling helpless, and even stranded, particularly in contrast with seemingly better cared-for Europeans. For instance, Cristina Pratt, who was visiting Morocco from the East Bay in California was quoted as stating "France is being very open with the citizens and is moving mountains to get them home; meanwhile the U.S. embassy says 'call the airlines' and 'prepare to be here for a while, but not indefinitely'" (Mzezewa, 2020a, 2020b, Mar. 19). Indeed, the German foreign minister was cited as promising to "do everything possible" to enable stranded Germans to return home (Stevis-Gridneff & Pérez-Peña, 2020, Mar. 17), while the Canadian prime minister acknowledged that, realistically, some of the "three million Canadians at any given moment around the world, living and working" would not be returning (Perreaux et al., 2020, Mar. 18).

The pandemic emerged within our twenty-first-century context of complex mobility patterns and, indeed, complex citizenship positions. While some of the accounts in our corpus highlighted complex positionalities, these were generally overlooked. By contrast, those with unambiguous mobility capital (Sheller, 2014) were engaged with, via the media, to prepare them for a period of 'forced' immobility. For instance, Julianna Strickland, Los Angeles-based founder of Space Camp Organizing, was quoted as offering the following advice to immobile tourists: "[t]reating yourself well at home means making a space that you want to be in, and setting yourself up for success as you head into whatever you're doing between trips" (Firshein, 2020, Mar. 31). Thus, the privilege of mobility capital—of enjoying both a positive relationship with one's home and the 'right', unambiguous citizenship—was portrayed, throughout our corpus, as a crucial characteristic of the tourist-subject.

(Not) Handling a Global Public Health Crisis

Up to 2020, the global tourism industry had been an increasingly key economic sector, a truism that is made abundantly clear in our corpus of media reports. As concern for public health was becoming a priority in jurisdictions worldwide, officials and other tourism industry stakeholders expressed particular concern for the wellbeing of the tourism market. In relation to the commercial aviation sector, NYT reporters Gelles and Chokshi (2020, Mar. 5) proposed that, "[t]he more fundamental issue posed by the coronavirus—that large swaths of the traveling public may simply stay off planes for the foreseeable future—is a far greater threat. [...] Commercial aviation, like the internet, is part of the connective tissue of the global economy". Although the authoritative voice of the World Tourism Organization (UNWTO) is notably absent from our data, the international agency released numerous notices and updates during the period of the pandemic's becoming, insisting that any response to COVID-19 should be "proportionate to the public health threat and based on local risk assessment" (UNWTO, Mar. 6; Mar. 11). Thus, the progressive and piecemeal implementation of a slew of public health measures such as increased sanitation and mask-wearing, social distancing and lockdown, and evacuation and repatriation may be seen as strategic to preserving tourism and travel activities. In extreme cases, the threat to public health was obscured and even blatantly underestimated. For instance, while tourists were quarantined aboard a cruise ship nearby and a 60-year-old German tourist died, Egypt's tourism minister, Khaled el-Enany, is reported to have "visited one of Luxor's temples [to] declare that it was safe and open to visitors [while] Egypt's health minister, Hala Zayed, warned people not to 'exaggerate' about the scale of the crisis" (Walsh, 2020, Mar. 9).

While our corpus of texts showcases various instances of crisis management at tourist destinations—including the Diamond Princess, Macau's gaming district, and Luxor, Egypt, to name but a few—it demonstrates the general lack of global cooperation and coordination in handling the coronavirus outbreak. For example, after the Westerdam cruise ship had been repeatedly turned away from ports, Cambodia allowed it to dock, a move for which it was lauded. The SCMP reported remarks by WHO chief Tedros Adhanom Ghebreyesus to the effect that, "[t]his is an example of international solidarity we have been consistently calling for [...] Outbreaks can bring out the best and the worst in people" (Reuters, 2020, Feb. 13). However, Cambodia's impetus for granting haven to the ship was questionable, with prime minister Hun Sen commenting that "our current disease around the world is fear and discrimination", and documented as "shaking hands with passengers and handing out roses". In addition, "all passengers were given free visas" (Reuters & DPA, 2020, Feb. 15). In general, the threat to public health was overshadowed by the threat to the tourism industry, while in the Westerdam case, solidarity was demonstrably leveraged for political purposes. During the same timeframe, a range of promotional campaigns and cancellation policies were rapidly implemented in an effort to stimulate demand for tourism and travel.

'COERCION' OF PRIVILEGED MOBILITIES?

Our selection of media texts points up the tension and negotiation between public health, economic growth, and the right to travel. The ambiguity regarding travel was particularly salient in the days leading up to and immediately following the announcement of the pandemic, especially in North America. More specifically, responsibility for evaluating the relative threat of infection was largely passed on to travellers, perhaps with a view to honouring their right to mobility (Bianchi et al., 2020).

In this regard, Kasara Barto, public relations manager at Squaremouth, a travel insurance comparison service, was quoted as advising: "If you're not going to enjoy your vacation because you're nervous about this, then it's probably not worth you going" (Caron, 2020, Mar. 17). However, this individualization of cost and benefit assessments further obscures the—earlier discussed—uneven impacts on collectives due to social and spatial conditions. Indeed, Alison Essom, who worked for an insurance company in Edinburgh, expressed an individualistic perspective on the threat to public health: "We're taking a calculated risk being here [...] But we're used to travelling where things aren't perfect. In Asia, we have to get rabies jabs" (Reguly, 2020, Mar. 6). In support of this kind of self-responsibilization in the face of a health risk, a slew of 'expert' advice and guidelines regarding sanitation, mask-wearing, and other protective tactics emerged in the media at this time.

For example, Corey Higgins, a mother of two who was visiting Disney World with her husband and children was quoted as saying, "We have wipes, we have hand sanitizer, we aren't elderly and honestly, if you're going to get it, you're going to get it. If I'm going to get sick and die, I might as well do it at Disney World" (Mzezewa, 2020a, 2020b, Mar. 12). Furthermore, the sunk cost of cancelling vacations that would not be fully refunded may have offered an additional incentive to travel. To paraphrase G&M reporter McGinn (Mar. 11), Mr. Pearce, who will be going on a week-long cruise next month, enjoys peace of mind knowing that, for someone like him in his late 30 s, the likelihood of dying from COVID-19 is very low. Moreover, quarantining is something that he is prepared to deal with: "If the cost is nothing and all it is is a time thing ... I'm not really all that concerned [...] You gotta relax. You gotta go on vacation throughout the year". Other travellers, including 30-year-old Joe DeSimone, took advantage of reduced prices: "The way I see it, [...] either things will normalize and prices will return to a rate that makes it difficult for me to travel, or the world is going to end and I might as well enjoy it while it lasts" (Wolfe, 2020, Mar. 13).

Similarly, Kelly McPhee, 31, a bartender working two jobs in Chicago found travel to be accessible for the first time in two years. "She's not worried about the virus, 'partly because I'm young, and partly because we've gone through this before. Ebola. Swine flu. Y2K. It just seems like the next thing'" (Wolfe, 2020, Mar. 13). With some media reports relaying messages dismissing the severity of the threat to public health, and others sharing individualized travel advice and promotions, we find that individuals were subtly coerced into remaining or becoming mobile while the pandemic was coming into being. Through the provision of guides to self-responsibilization and individual risk assessment, as well as through marketing strategies appealing to those for whom travel is not always accessible, we see the freedom to travel presented as a false sense of free will (Korstanje, 2018) in the context of an emergent global public health crisis.

WAITING FOR DEATH

The retreat of urban residents and remote workers to second homes, vacation rentals, and nonurban tourism destinations was also well documented by the media, especially following the becoming of the pandemic. In our corpus, the figure of the 'digital nomad' was not explicitly evoked by reporters, although we get the sense that novel privileged pandemic (im)mobilities were on the horizon. In the weeks, months, and even years following 11 March 2020, digital nomadism became a prominent focus for the travel industry, fuelling novel marketing strategies, border infrastructures (e.g., special visas) and, indeed, scientific inquiry (e.g., Holleran & Notting, 2023). Nevertheless, evidence of these novel urban-to-rural flows was already reported in East Asia during the becoming of the pandemic phase, with an emphasis on the relative vulnerability of local peoples and places in relation to competition for limited resources and exposure to the coronavirus. Xi Fengxian, a hotelier in Ejina Banner, Inner Mongolia in the Gobi Desert was quoted as stating "… it's inevitable that people will flee here […] And given the extremely poor medical infrastructure here, once we are infected, all we will be able to do is wait for death" (Vanderklippe, 2020, Feb. 25). Although beyond the scope of our analysis here, this 'inevitability' of privileged mobilities raises questions about the right to free (urban to rural) movement and the potential consequences for any individual or state actor who may infringe upon that right. Given the scope and scale of urban-to-rural flows, the reconfiguration of living and travelling practices following the implementation of public health measures worldwide, and the impacts of these flows on residents and communities, this biopolitical trend merits additional scientific attention in its own right.

The Biopolitical Becoming of Tourism

As Agamben (1998) and many other political theorists have pointed out, around the late nineteenth century the modern liberal state moved from governing territories to governing populations. Tourism, with the tourist as a transient post-political subject (Ek & Tesfahuney, 2016), may be seen as a manifestation of this shift in the focus of power from land to bodies. However, during the pandemic, the mass repatriation of foreign nationals around the world, followed by the closure of national borders, and even the creation and enforcement of subnational borders, represented an unexpected return to territorial power-wielding by nation-states. Policy responses also entailed the reaffirmation of biopower within national boundaries, via lockdowns, the blocking of interregional mobility, and, eventually, vaccine mandates. While the COVID-19 pandemic brought tourism to an abrupt halt in 2020, it also exposed its biopolitical dimensions.

First came the acknowledgement of bare life[2] as a threat to all life, which justified the implementation of a state of exception in an effort to contain the spread of the novel coronavirus via the mobile bodies of tourists. The docking of the Diamond Princess in the harbour of Yokohama, Japan, was the first and best-documented case of such a state and space of exception. Once the vessel was placed under quarantine, the rules inherently in force on board the cruise ship changed, as did the interaction between the ship and the outside world. This state of exception was eventually to be extended to most of the world, after the disease outbreak was attributed pandemic status. Thus, the Diamond Princess represented a sort of laboratory of what was to come. Indeed, the state of exception reordered society according to new criteria (i.e., essential/non-essential; social distance/proximity) as initially witnessed on a smaller scale in the cruise sector. Given that the more affluent passengers had access to better cabins, their personal accounts describe a situation that was an extension of their holidays, with stress due to the virus threat mitigated by the comfort of luxury accommodation. In contrast, other passengers were confined to cabins that were not suited to a prolonged quarantine. At

---

[2] In Agamben's (1998) work, bare life refers to the sheer biological fact of life, to which power reduces human life in order to manage, control, and punish it. He gives the example of the Guantanamo Bay prisoners, who were stripped of their political, legal, and social status and reduced to their bare life.

the same time, the staff of the ship were mainly invisibilized and at risk of contracting the disease while performing 'essential tasks' as productive bodies. This reordering of bodies according to the relative comfort of one's quarantining arrangements and role in 'essential activities' would eventually expand worldwide as the pandemic-driven state of exception became generalized (Rose, 2021).

While the pandemic threatened both life and capital, it also created an intersection between threatening bodies as vectors of the virus and the productive bodies of capitalism. For instance, we earlier highlighted tourists' accounts of returning home while being let down by the very technological and political apparatuses that had initially enabled them to travel. The possibility to come home and to choose where to be immobile meant that individuals could secure their bare lives within safe and comfortable spaces, where they would also be able to maintain their social lives thanks to remote working arrangements, digital communications, and inhabiting low density communities, enacting political life thanks to physical distance. However, confusion surrounding what, or where, home was for individuals working, studying, living, or simply 'stranded' abroad illuminated the complex interaction between territory and population. Indeed, as the borders began to harden, we witnessed differentiation in the conditions of exception from nation-state to nation-state: the borders of the body politic were redefined, and by extension, the touristic body was momentarily dismembered. This was initially manifested when different countries brought forth different rules and discourses regarding the docking of cruise ships. While the Japanese authorities were creating a highly enclosed space of exception vis-à-vis the Diamond Princess, Cambodia was implementing other exceptions by welcoming the Westerdam's passengers and fast-tracking their visas at no cost.

As previously noted, workers' experiences seldom featured in our corpus on the becoming of the pandemic. Indeed, at that time, tourism employees were mainly cited when reiterating corporate positions and policies in relation to the spread of the virus. However, the lead-up to the pandemic placed workers in a difficult bind. As productive bodies, they needed to go on working despite representing a threat of contagion for tourists, the consuming bodies. This bind raised questions about whose life to protect, whose life to sacrifice, whose life to save, and whose life to abandon. Indeed, tourism itself, with its global economic footprint, was facing the threat of non-circulation. Therefore, as our corpus shows, the total halting of tourism could not be contemplated at first, an eventuality that was eclipsed by a range of narratives produced to justify the continuance of international travel. Some bodies needed to remain in circulation to maintain the population/production dyad.

These narratives encompassed: the setting up of tentative virus containment zones (i.e., the Diamond Princess); a 'responsibility narrative' (Tremblay-Huet & Lapointe, 2021), which called for measured responses because tourism-dependant areas needed to keep tourism going; and finally, an 'immunization narrative' surrounding tourists, whose purpose was to keep the tourism economy alive. Accordingly, individual tourists internalized sanitary rules in their discourses to minimize the risk of getting infected, while muting the possibility that they might infect others. The consuming bodies of tourists 'wielded' hand sanitizer and masks, and young people asserted their desire to travel, to the point of deeming it better to die travelling than at home. This exposes, we would argue, the importance of tourism as a component of the individual neoliberal subject, who claimed tourism as an individual right while economic stakeholders supported discourses of least infringement and fast recovery via a responsibility narrative. Within the population/production dyad, the tourism industry accepted mobile subjects' spending and consuming while desubjectifying exposed workers, who would eventually disappear behind personal protection equipment as the pandemic became generalized. Indeed, differential immunity discourses were deployed to justify each of these two patterns: while tourists were internalizing sanitary measures in order to keep on travelling, a high risk of infection was imposed on workers in order to immunize the economy against the collapse of tourism. Thus, discipline was imposed on the workers so that the 'kinetic elite' might be kept relatively safe, enabling them to retain their privileged status and, thus, maintaining the circulation of consumers' bodies and capital.

Although the years-long period of lockdown and vaccine mandates is now behind us, the biopolitical processes that were set in motion may still be unfolding. Indeed, their long-term impacts are difficult to pin down, given that new and ancient crises act with and through them (Cheer et al., 2021). This analysis of the pandemic's becoming phase in early 2020 points up multiple—social, spatial, and political—emergent tendencies. These include the internalization by individual subjects of health and safety rules designed to mitigate health risks and justify a return to high-intensity leisure mobility, together with the (re)emergence of rural areas as productive spaces for contemporary de-territorialized experiential capitalism and a return to the debate on essential versus non-essential mobilities. These perspectives, which our hybrid discourse and narrative analysis of mass media reports has brought into focus, point to broader ongoing biopolitical dimensions of tourism.

Competing Interests

The authors have no conflicts of interest to declare that are relevant to the content of this chapter.

References

Agamben, G. (1998). *Homo sacer*. Stanford University Press.

Arellano, A., & Nasab, P. S. (2020). Eaux troubles: Les navires de croisière au temps de la COVID-19. *Téoros. Revue de recherche en tourisme, 39–3*. Retrieved June 30, 2023, from http://journals.openedition.org/teoros/7548

Bauman, Z. (1996). Tourists and vagabonds: Heroes and victims of postmodernity. *Political Science Series, 30*. Wien Institute for Advanced Studies, Vienna. Retrieved June 30, 2023, from https://irihs.ihs.ac.at/id/eprint/898/1/pw_30.pdf

Bianchi, R. V., Stephenson, M. L., & Hannam, K. (2020). The contradictory politics of the right to travel: Mobilities, borders & tourism. *Mobilities, 15*(2), 290–306. https://doi.org/10.1080/17450101.2020.1723251

Brouder, P., Teoh, S., Salazar, N. B., Mostafanezhad, M., Pung, J. M., Lapointe, D., Higgins Desbiolles, F., Haywood, M., & Hall., M.C., & Clausen, H. B. (2020). Reflections and discussions: Tourism matters in the new normal post COVID-19. *Tourism Geographies, 22*(3), 735–746.

Brown, W. (2015). *Undoing the demos: Neoliberalism's stealth revolution*. Mit Press.

Campbell, T. C., & Sitze, A. (Eds.). (2013). *Biopolitics: A reader*. Duke University Press.

Cheer, J. M., Lapointe, D., Mostafanezhad, M., & Jamal, T. (2021). Global tourism in crisis: Conceptual frameworks for research and practice. *Journal of Tourism Futures, 7*(3), 278–294. https://doi.org/10.1108/JTF-09-2021-227

Ek, R., & Hultman, J. (2008). Sticky landscapes and smooth experiences: The biopower of tourism mobilities in the Öresund Region. *Mobilities, 3*(2), 223–242. https://doi.org/10.1080/17450100802095312

Ek, R., & Tesfahuney, M. (2016). The paradigmatic tourist. In *Tourism research paradigms: Critical and emergent knowledges*. Tourism Social Science Series, Vol. 22. Emerald Group Publishing Limited, pp. 113–129

Ek, R., & Tesfahuney, M. (2019). Topologies of tourism enclaves. *Tourism Geographies, 21*(5), 864–880.

Esposito, R. (2011). *Immunitas: The protection and negation of life*. Polity Press.

Foucault, M. (1997). *Il faut défendre la société, Cours au Collège de France (1975–1976)*. Gallimard Seuil.

Foucault, M. (2004). *Naissance de la biopolitique, Cours au Collège de France (1978–1979)*. Gallimard Seuil.

Foucault, M. (2007). *Security, territory, population: Lectures at the Collège de France, 1977–1978*. Palgrave Macmillan.

Holleran, M., & Notting, M. (2023). Mobility guilt: Digital nomads and COVID-19. *Tourism Geographies, 1–18,*. https://doi.org/10.1080/146 16688.2023.2217538

Kadri, B., Lapointe, D., & Tacherifet, S. (2023). Rethinking or reinventing tourism? Exposing the ontological and epistemological conflicts in tourism studies literature during the COVID-19 pandemic in *Tourism & Recreation Research*. https://doi.org/10.1080/02508281.2023.2224705

Korstanje, M. E. (2018). *The mobilities paradox: A critical analysis*. Edward Elgar.

Lapointe, D. (2022). Dispositif de contrôle, état d'exception et corps touristiques : le tourisme comme phénomène biopolitique. *Via Tourism Review, 21*. Retrieved June 30, 2023, from http://journals.openedition.org/viatou rism/8065

Minca, C. (2010). The Island: Work, tourism and the biopolitical. *Tourist Studies, 9*, 88–108. https://doi.org/10.1177/1468797609360599

Minca, C. (2015). The biopolitical imperative. In J. Agnew, V. Mamadouh, J. Sharp, & A. J. Secor (Eds.), *The Wiley Blackwell companion to political geography* (pp. 165–186). John Wiley & Sons.

Renaud, L. (2020). Reconsidering global mobility—distancing from mass cruise tourism in the aftermath of COVID-19. *Tourism Geographies, 22*(3), 679–689. https://doi.org/10.1080/14616688.2020.1762116

Rose, J. (2021). Biopolitics, essential labor, and the political-economic crises of COVID-19. *Leisure Sciences, 43*(1–2), 211–217. https://doi.org/10.1080/01490400.2020.1774004

Sarrasin, B., & Lapointe, D. (2021). Le tourisme au temps de la Covid-19 : Entre géopolitique et biopolitique. *Diplomatie, 110*, 80–82. Retrieved June 30, 2023, from https://www.areion24.news/2021/11/15/le-tourisme-au-temps-de-la-covid-19-entre-geopolitique-et-biopolitique/

Sheller, M. (2014). The new mobilities paradigm for a live sociology. *Current Sociology, 62*(6), 789–811.

Sheller, M. (2016). Uneven mobility futures: A foucauldian approach. *Mobilities, 11*(1), 15–31. https://doi.org/10.1080/17450101.2015.1097038

Tremblay-Huet, S., & Lapointe, D. (2021). The new responsible tourism paradigm: The UNWTO's discourse following the spread of COVID-19. *Tourism and Hospitality, 2*(2), 248–260. https://doi.org/10.3390/tourho sp2020015

United Nations World Tourism Organization (UNWTO). (2020, March 6). Tourism and coronavirus disease (COVID-19), Tourism resilience. *World Tourism Organization*. Retrieved June 30, 2023, from https://webunwto.s3. eu-west-1.amazonaws.com/s3fs-public/2020-03/Tourism-COVID19_EN. pdf

United Nations World Tourism Organization (UNWTO). (2020, March 11). UNWTO and WHO agree to further cooperation in COVID-19 response. *World Tourism Organization*. Retrieved June 30, 2023, from https://www.unwto.org/unwto-and-who-agree-to-further-cooperation-in-covid-19-response

Textual data

Associated Press. (2020, February 7). Coronavirus: how a luxury cruise became a 'floating prison' during quarantine in Japan. *SCMP*.

Caron, C. (2020, March 17). Coronavirus is spreading. Should you cancel your vacation? *The New York Times*.

Firshein, S. (2020, March 31). How to store your travel gear. *The New York Times*.

Gelles, D., & Chokshi, N. (2020, March 5). 'Almost without precedent': Airlines hit hard by coronavirus. *The New York Times*.

The Globe and Mail. (2020, February 13). Passengers uncertain but hopeful as cruise ship still under Japanese quarantine. *The Globe and Mail*.

Larsen, R. (2020, March 24). How to see the world when you're stuck at home. *The New York Times*.

McGinn, D. (2020, March 11). Canadians debate cancelling cruises amid travel warnings due to coronavirus outbreak; The fear of what might happen on a cruise has left many Canadians scrambling to either get a refund or reschedule cruises they already have booked. *The Globe and Mail*.

Mzezewa, T. (2020, March 12). 'If I'm going to get sick and die, I might as well do it at Disney World'. *The New York Times*.

Mzezewa, T. (2020, March 19). Many feel 'abandoned' by U.S., far from home. *The New York Times*.

Perreaux, L., Hayes, M., & MacKinnon, M. (2020, March 18). Canadian travellers stranded abroad after COVID-19 closes borders; From Peru to Ukraine, a growing number of Canadians seek a way home vis a vis cancelled flights and border closures. *The Globe and Mail*.

Reguly, E. (2020, March 6) A postcard from outbreak-stricken Italy. *The Globe and Mail*.

Reuters. (2020, February 13). No Westerdam cruise ship passengers test positive for coronavirus after arriving in Cambodia. *SCMP*.

Reuters & DPA. (2020, February 15). Coronavirus: passengers celebrate 'best cruise ever', after Westerdam ship docks at Cambodia. *SCMP*.

Rich, M. (2020, February 22). We're in a petri dish: How coronavirus ravaged a cruise ship. *The New York Times*.

Rich, M. & Wong, E. (2020, February 17). They escaped an infected ship, but the flight home was no haven. *The New York Times*.

Stevis-Gridneff, M., & Pérez-Peña, R. (2020, March 17). Europe barricades borders to slow coronavirus. *The New York Times*.

Vanderklippe, N. (2020, February 5). Coronavirus lockdown spreads to China's remote areas: Some of the most rigid response measures are in small communities, where roads are closed. *The Globe and Mail*.

Walsh, D. (2020, March 9). In Egypt, tourists torn between ancient temples and coronavirus tests. *The New York Times*.

Wolfe, J. (2020, March 13). Young, confident and flying, virus be damned. *The New York Times*.

CHAPTER 4

Re-habituation and the More-than-human Biopolitics of Gorilla Tourism in Uganda

Amos Ochieng◉*, Christine Ampumuza*◉*,
and Maartje Roelofsen*◉

Abstract 'Habituation' is often described within conservation and tourism studies as a relational process by which animals and humans become accustomed to one another's bodily presence. This chapter offers a biopolitical reading of habituation within the context of gorilla tourism

A. Ochieng
Department of Forestry, Biodiversity and Tourism, School of Forestry, Environmental and Geographical Sciences, Makerere University, Kampala, Uganda
e-mail: amos.ochieng@mak.ac.ug

C. Ampumuza
Department of Tourism and Hospitality, Kabale University, Kabale, Uganda
e-mail: champumuza@kab.ac.ug

M. Roelofsen (✉)
Open University of Catalonia, Barcelona, Spain
e-mail: mroelofsen@uoc.edu; maartje.roelofsen@wur.nl

Department of Economics and Business, Universitat Oberta de Catalunya, Barcelona, Spain

M. Roelofsen and C. Minca (eds.), *Tourism and Biopolitics in Pandemic Times*, https://doi.org/10.1007/978-3-031-46399-0_4

in *Bwindi Impenetrable National Park* in Uganda during the COVID-19 pandemic. Employing ethnographic methods and desk-based research, we examine the implementation of new biosecurity controls and interventions for both humans and animals throughout 2020–2023, drawing on more-than-human biopolitical theory and relational approaches to animal agency in tourism. The findings show that the pandemic has exacerbated existing inequities that had historically been sustained by conservation biopolitics, affecting both human and non-human species. The chapter also raises questions about the ongoing efforts to (re)habituate gorillas, and not merely within the context of crisis. Given the ephemeral nature of tourism visits, what do tourists contribute to sustaining the human-animal relations for the benefit of both human and non-human species?

Keywords Habituation · More-than-human biopolitics · Gorilla tourism · Uganda · COVID-19 pandemic

INTRODUCTION

Gorilla tourism at *Bwindi Impenetrable National Park* (hereafter Bwindi) in Uganda, which has been ongoing since the early 1990s, is organized around the notion of 'habituation', often described within conservation and tourism studies as a relational process by which animals and humans become accustomed to one another's bodily presence (Ampumuza & Driessen, 2021). Among other purposes, habituation allows animals such as gorillas to become 'viewable' as subjects/objects of tourism consumption and scientific study, which arguably benefits local communities and the tourism industry in terms of the related economic rewards (Samuni et al., 2014). Habituation can take from a few days to several years to achieve its 'desirable' effects, that is to say, for an animal to lose its fear of humans and, relatedly, to perceive the presence of human beings as 'neutral' (Doran-Sheehy et al., 2007). In the context of Bwindi, different groups of humans have historically been involved in the habituation of gorillas, including scientists, trackers, local communities, rangers, and

Cultural Geography Group, Wageningen University and Research, Wageningen, The Netherlands (indirectly) tourists. These subjects track and map gorilla trail networks and nesting areas, approach gorillas within a certain distance, visibly position their bodies vis-à-vis those of gorillas, and signal their presence by vocalizing softly (ibid.). They observe and mimic the gorillas' bodily behaviours—such as sitting, picking and eating leaves, and scratching—in the attempt to be perceived as non-threatening (Ampumuza & Driessen, 2021). During habituation, collecting behavioural, demographic, and spatial data about gorilla life is of vital importance to 'conservation biopolitics'; this includes tracking the primate's daily trajectories, measuring their levels of aggression and time spent in contact, as well as recording births and deaths and many other types of information. Within conservation biopolitics, ecology has become "an important power/knowledge regime inside and outside protected areas" which shapes "how nature should be understood" and whose experts are "in charge of drawing distinctions between what is (not) natural, and where nature should begin and end" (Bluwstein, 2018, p. 152).

National parks in Africa have been flagged by Bluwstein as colonial biopolitical spaces par excellence; specifically, they demonstrate how preservation and conservation rationalities translate into the governance of human and animal populations, such that certain lives must be protected (let live) while others are abandoned (let die) (Bluwstein, 2018). In the case of Bwindi, as with other national parks (Ampumuza & Driessen, 2021), habituation is part of a broader set of wildlife and conservation measures and practices that seek to protect and manage gorilla populations within a specific territory. Although Bwindi had already become a Reserve under the British colonial government in 1932, the *Game Act* that placed restrictions on hunting in the park—mainly with a view to protecting the endangered mountain gorillas—only came into effect in 1964. Since the colonial period, sub-Saharan Africa has seen multiple biopolitical strategies for wildlife preservation and conservation and human development, whereby "people and wildlife became classified, categorized and essentialized into populations with distinct properties and clearly delineated territories" (ibid., p. 146). These strategies include the classification of human populations, such as the indigenous Batwa people, who had always lived side by side with gorillas in Bwindi until the latter became a protected species.

In this chapter, we offer a biopolitical reading of habituation within the context of managing gorilla tourism and animal life in Bwindi during the COVID-19 pandemic. The outbreak of the pandemic and the related halting of tourism mobility had major implications for gorilla–tourist interaction and the related habituation practices in Bwindi. Drawing on both ethnographic methods and desk-based research, we examine the implementation of new biosecurity controls and interventions for both humans and animals throughout 2020–2023, drawing on more-than-human biopolitical theory and relational approaches to animal agency in tourism.

More-than-human Biopolitics and Tourism Studies

While the introductory chapter in this edited volume discusses the emergence of a 'biopolitical turn' in tourism studies, here we wish to zoom in on the anthropocentric nature of this shift in tourism scholarship and advocate for a *more-than-human approach* to the politics of life. This is in recognition of the fact that while for decades a diverse body of tourism research has been focused on human–animal relations, including in the context of gorilla tourism (van der Duim et al., 2014), much of this work has been characterized by an emphasis on ensuring that human-centred socio-economic goals may be attained (Srinivasan, 2017). Examples are fortress and community-based conservation models (Ampumuza, 2022), neoliberal conservation models such as market-based approaches (McAfee, 1999) and, more recently, convivial conservation approaches (Büscher & Fletcher, 2019; Ochieng et al., 2023). In the view of Ampumuza and Driessen, such (extreme) conservation models embody "a form of biopolitics that not only combines care with intervention but permits intrusion into the lives and bodies of individual animals in the name of species conservation" (2021, p. 1604). These models also reinforce the continued separation of the governance of natural landscapes from the governance of people (Bluwstein, 2018; Braun, 2013; Büscher & Fletcher, 2019; Srinivasan, 2017).

The fields of environmental and animal studies already offer well-consolidated critiques of anthropocentric understandings of biopolitics and biopower, including in relation to the work of Foucault, Agamben, and Hardt and Negri (Chrulew, 2012; Chrulew & Wadiwel, 2017). While Foucault acknowledged that "animalisation" is used as a political means of "rationalising violence against various marginalised groups in human societies", he never paid much attention to actual animals in his work (Chrulew, 2017, p. 4). Nor did he challenge "the logic of speciesism", which enables a "matrix of oppression" on the part of *all life*, including the various forms of power and violence that are exerted upon nonhuman life (Chrulew, 2012, p. 54). Despite his apparent interest in the history of biology and the biologization of human politics, Foucault never thematized "human relations with nonhuman animals in a way that politicised [animals'] subjected bodies and lives" (Chrulew & Wadiwel, 2017, p. 4).

Agamben, on the other hand, has written at length about the distinction between the human and nonhuman and between humanity and animality in the tradition of Western political thought, including in his works *The Open* (2004) and *Homo Sacer* (1998). In his discussion of the "anthropological machine", Agamben makes frequent critical references to mythical animal figures to exemplify the *passage* or *threshold* between humans and animals (2004). The most famous instance of this is the werewolf, a threshold figure that is banned by his [sic] community from participation and that moves between the city and forest (Minca, 2006, p. 392). The werewolf is "a subject torn not only between life and death, but also between place and its 'measure' (space)—excluded by both but, at the same time, constitutive of both" (ibid.). Crucially, this mythical figure serves as a *metaphor* for exemplifying the fundamental indistinction between human and animal. Nevertheless, Agamben "thematises the human/animal distinction as a central site for the production of human subjectivity, yet thematises the political effects of this caesura only on human subjects" (Chrulew & Wadiwel, 2017, p. 9). He also pays little consideration to how nonhuman animal lives are made *bare*, "reduced through an abject status, both epistemologically and ontologically" (Chrulew, 2012, p. 58), be it via millennia of extensive breeding or subjection to mechanized factory farming and industrialized slaughter technologies (see for example, Wadiwel, 2018).

Hardt and Negri's conceptualization of biopolitics and biopower assumes a new world order that they have termed 'Empire'. These authors conceive Empire as a new stage in global capitalism, which is marked by the integration *of* and connection *between* all states and regions of the world. This form of capitalism relies on different forms of production with respect to industrial capitalism: it is highly informatized and networked, and is not only sustained by physical labour but also by various forms of immaterial labour that "involve[...] the production and manipulation of affects and require[...] (virtual or actual) human contact, labour in the bodily mode" (Hardt & Negri, 2000, p. 293). In this new world order and the related understanding of biopolitics, the whole of society is subsumed under capitalism and assumes utility and value, as does *nature*.

But rather than being merely exploited and regulated, nature is now also 'machine-made'. In his interpretation of Hardt and Negri's work on biopolitics, Thomas Lemke argues that forms of life such as "biological and genetic diversity" translate into economic growth, via the convergence of economics, nature, politics, and culture (Lemke, 2011, p. 70). Thus, animals on the verge of extinction (e.g., mountain gorillas) may be converted into profitable forms of life within the context of 'eco-tourism' or 'sustainable tourism'.

In a departure from Agamben's negative understanding of biopolitics and biopower, Hardt and Negri see ample potential for social and political resistance against biopower on part of the heterogeneous set of actors that make up the "multitude" (Hardt & Negri, 2004, p. 66). The multitude, they argue, are subjectivities "capable of creating a new world", predominantly via the transformative potential of their immaterial labour, which produces ideas, knowledges, and affects, which in their turn also affect other forms of labour (ibid., p. 65). For Hardt and Negri, "immaterial labor is biopolitical in that it is oriented toward the creation of forms of social life"; it produces and reproduces society as a whole (ibid., p. 66). Critics have viewed Hardt and Negri's work as open to the inclusion of concern for the animal realm. In *Multitude*, for example, the two authors argue that (some) human–animal relations have always been fundamental to life and do not translate into economic wealth. Indigenous populations in the Amazon (and other communities in the Global South), they maintain, have lived-with and kept-alive plants and animals of different species to keep themselves alive and profit from their beneficial qualities (ibid., pp. 131–132). However, when Hardt and Negri posit that the multitude is characterized by "innumerable internal differences"—a multiplicity of singular differences—such as "different cultures, races, ethnicities, genders, and sexual orientations", they omit to mention *species* as a category of difference, and therefore exclude "nonhuman animals from their conception of the human multitude" (Chrulew, 2012, p. 61). Furthermore, in their discussion of labour, only human labour is seen as productive, despite humans' vital reliance on, and gain from, the physical and affective labour of animals (ibid.; Barua, 2017). Think—for example—of bees' production of honey, or zoo and circus animals' contribution to eliciting emotions among visitors/audiences. Similarly, although there are innumerable forms of nonhuman intelligence and organization, productivity, communication and self-constitution, Hardt and Negri's account excludes animals from the creative potential of the multitude (ibid.).

Despite the shortcomings in the key works on biopolitics and biopower that we have just reviewed, biopolitical theories have nonetheless been fruitfully applied in research that focuses on nonhuman animal life to engage with a wide range of topics (for an overview, see Chrulew & Wadiwel, 2017, pp. 6–8). This chapter is informed by 'conservation biopolitics', a subdomain of biopolitics that encompasses endangered species management, conservation breeding and genetics, protected areas, and rewilding. These four biopolitical domains have all been produced and sustained by apparatuses of knowledge and treatment concerning animals and they rest on "sets of scientific practices and discursive understandings [that] produce valuations of life that appear natural, universal, and technically derived even as they are particular and normative" (Biermann & Anderson, 2017, p. 2).

The politics of animal and wildlife conservation has long been a key concern of tourism scholars. It is a politics that fundamentally departs from the understanding that certain animals should be let live while other animals should be let die, and that is maintained via a diversity of apparatuses (as Foucault might term them) spanning institutions, scientific research, (inter)national legislation, policies, and practices. Such is also the case with the conservation of critically endangered nonhuman primate species, which extends beyond the lives of gorillas. For example, conservation efforts and related tourism niches have sprung up around orangutans in Indonesia and Malaysia, and around lemurs in Madagascar, to name but a couple (Russon & Susilo, 2014; Wright et al., 2014). Yet other nonhuman primates have been excluded from protection and commoditization, in the sense that they are not viewed as 'endangered' enough or, alternatively, are perceived and treated as pests or problem animals, a fate that has sometimes befallen baboons, even while they have been targeted for conservation in other regions (Strum & Nightingale, 2014).

In the next sections, we explore the changing role that preservation and conservation rationalities and initiatives have played in governing humans and animals (especially gorillas) in Bwindi as a function of the pandemic biopolitics of 'making live' and 'letting die'.

Techniques of Wildlife Conservation
and Biopolitical Governance at Bwindi

The potential of biopolitics as a productive analytical framework for wildlife and nature conservation has been extensively described in previous research focused on different (post-) colonial contexts (see for example, Bluwstein, 2018, and Mokhele, 2022). The deployment of biopolitical strategies to separate distinct populations of humans and animals from another—in an attempt to protect, govern, and render productive certain forms of life as opposed to others—is similarly an integral part of Uganda's colonial history. As noted above, Bwindi was first declared a 'reserve' by the British colonial government in 1932, a decision informed by a protectionist philosophy and a Western rationality that implemented a nature-culture divide by means of spatial separation . By separating humans from nonhuman–animal populations (such as gorillas) through the imposition of (forest) boundaries, these two groups could arguably be protected from one another. However, according to Blomley (2003, p. 233), Bwindi's 'reserve' status initially continued to entail "relatively liberal (and rarely enforced) regulations regarding access rights" for forest-dependent human households. Other areas of Uganda that were viewed as abounding in wildlife and perceived as facing similar threats from human existence were also conferred with the new status of reserves (Ampumuza, 2022; Butynski & Kalina, 1993). By the 1980s, the mountain gorilla population in Bwindi was on the verge of extinction due to poaching and habitat destruction, with about 620 individuals remaining in the forest by 1989. Consequently, numerous conservation organizations, policy makers, and scientists advocated for the subspecies to be included on the International Union for Conservation of Nature (IUCN) Red List of Highly Threatened Species (see Hickey et al., 2018).

Given the deemed conservation importance of Bwindi, the reserve was declared a 'National Park' in 1991, with numerous (spatial) implications including new restrictions on any form of legal resource use afforded by the forest. The transition from 'Reserve' to 'National Park' affected both humans and animals. It meant, for example, that the original inhabitants of Bwindi—the Batwa people, among other indigenous groups—were evicted and resettled on the borders of the forest, and consequently themselves faced hardship, starvation, and extinction (Tumushabe & Musiime, 2006). It inadvertently also led to these groups' traditional means of subsistence such as hunting becoming redefined as an illicit poaching practice. Gorillas, on the other hand, continued to cross the established forest boundaries as before, foraging in areas that were designated for agricultural production and other human-oriented purposes (Ampumuza & Driessen, 2021). The Bwindi Impenetrable Forest Conservation Trust (hereafter BMCT, established in 1994), set out to deal with the consequences of the conservation interventions, arguing that "without healthy and economically secure communities around Bwindi and Mgahinga protected areas, the area remains under threat from grazing, hunting and such vices" (BMCT, 2023). According to BMCT's website, its aim is to improve 'the quality of life' of Bwindi's surrounding communities "by providing education, health services, safe water, vocational training and sustainable resource use skills in this impoverished and most densely populated region in Uganda" (ibid.). Hence, one of the BMCT's biopolitical goals continues to be the conservation of animal lives through the governance of human lives in and around Bwindi. This approach particularly concerns the 'social body' or biological corpus of the Batwa people, in relation to their health status, life span, and the production and circulation of wealth, as articulated on the BMCT website. Fostering and managing the existence of the Batwa people as a population is seemingly pivotal to the protection of the lives of more-than-human others; a form of biopower that appears as a *vital force*, as multiple others work through negative instances such as 'harm', 'deprivation', and 'extinction' to collectively create an affirmative alternative (Braidotti, 2015).

Bwindi's status as a conservation area was further legitimized and reinforced by other (international) conservation authorities, not least by UNESCO who added the region to its World Heritage List in 1994. According to UNESCO's website, Bwindi is "home to almost half of the world's mountain gorilla population, the property represents a conservation frontline as an isolated forest of outstanding biological richness surrounded by an agricultural landscape supporting one of the highest rural population densities in tropical Africa. Community benefits arising from the mountain gorilla and other ecotourism may be the only hope for the future conservation of this unique site" (UNESCO, 2023). On this dedicated webpage, references abound to desirable versus undesirable populations and related population management through 'boundary making': "the Park boundary is clearly delineated with planted trees and concrete pillars as markers along areas where rivers do not form the boundary. This clear boundary line has mostly stopped encroachment by the local communities, although with increasing population, agricultural encroachment will remain a potential threat" (UNESCO, 2023). In order to address potential conservation-versus-livelihood challenges in the area, the Ugandan government introduced a strategy of gorilla habituation in 1991 (Ampumuza & Driessen, 2021) and since 1993 gorilla tourism has been the main source of revenue for the park, with sales of 'gorilla viewing permits' topping US$ 31 M between 2005 to 2017 (Uganda Wildlife Authority, 2018). Even though revenue sharing schemes allocate specific percentages of this income to local communities, priority is given to gorilla conservation and the protection of the Bwindi Forest itself (Plumptre et al., 2003; Robbins et al., 2011; Weber et al., 2020).

Since becoming a protected species within the National Park space, the mountain gorilla population has displayed a pattern of recovery, with around 1063 individuals inhabiting the forest today. Because of this positive development, the IUCN has removed the gorillas from the Red List of *highly endangered* species. Now classified as 'merely' *endangered*, the gorillas continue to receive protection and remain subject to habituation (Mittermeier et al., 2022). As noted by Weber et al. (2020), there are multiple reasons to continue the habituation of gorillas at Bwindi. Not only does the rising gorilla population arguably need to remain accustomed to the presence of humans, but tourist demand to visit gorillas in Uganda has also increased over recent decades.

Having provided a brief genealogy of the biopolitical dimensions of gorilla conservation in Bwindi, we now move on to explore changes that have been observed in the park—in relation to the governance of humans and gorillas—since the outbreak of the COVID-19 pandemic. As earlier noted, we collected our data using ethnographic methods and an extensive desk-based review of both published and grey literature. We also base our findings on previous research conducted by the second author (Christine Ampumuza) on gorilla habituation in Bwindi, and specifically on her fieldwork in the Ruhija, Nkuringo, Rushaga, and Buhoma sectors (i.e., regions) of Bwindi with habituated gorilla families open to tourist visitation. In all, eighteen structured interviews were held with different officials, as well as with tour guides, rangers, and senior park staff involved in gorilla tourism in different capacities. All interviews were audio-recorded, transcribed, and analysed.

Pandemic-related Biopolitical Interventions and Biosecurity Regimes at Bwindi

The outbreak of the COVID-19 pandemic has offered a unique context for rethinking conservation (bio)politics and habituation at Bwindi from a more-than-human perspective. We would like to begin by advancing the critical observation that coronaviruses are *zoonotic* in nature, which means that they can be passed between humans and animals and thus cause infection and possible death in either species. Viruses that cross species barriers have traditionally led to the implementation of complex biosecurity regimes, including the mass culling of animals to contain disease outbreaks in relation to, for example, avian influenza or foot and mouth disease (see Hinchliffe & Bingham, 2008). Zoonotic viruses thus bear profound implications for how specific human and nonhuman–animal populations are valued and managed in relation to one another. They lay bare the divisions and hierarchies that are maintained between specific species, particularly when one population enjoys protective status while another does not. Since the onset of the pandemic, in the context of Bwindi specifically, biopower has continued to be directed towards fostering gorilla lives via the management of human mobility and related border regimes. Despite the unprecedented scope and impact of the COVID-19 pandemic globally, the threat to gorillas of human-transmitted disease is not new. In fact, biosecurity regimes have long been enforced around gorillas to ensure safe observation *of* and continued interaction *with* gorillas. This is also because one of the greatest recorded threats to gorilla lives is represented by human-transmitted infectious diseases; measles and respiratory infections have in the past proven to significantly impact gorilla mortality (Goldsmith, 2014). Multiple studies have shown that the more primates are exposed to tourists at close range, the higher their risk of contracting what are considered human-borne diseases (Köndgen et al., 2008; Sandbrook & Semple, 2006; Weber et al., 2020). Wearing facemasks was already a common practice among scientists working in gorilla conservation projects before the onset of the pandemic. Nevertheless, the COVID-19 pandemic also presented new and unique challenges to gorilla tourism, particularly because of the disease's highly virulent nature and its global spread via international (tourism) mobility.

Overall, the outbreak of the COVID-19 pandemic led to complex and differentiated responses to tourism in Uganda, including the banning of

all international travel for long periods, enforced lockdowns, the observance of standard operating procedures including social distancing and
wearing face masks, and mandatory vaccination as one of the requirements for the resumption of travel. On 23 March 2020, all airports
and borders in Uganda were closed to all passenger planes and vehicles,
meaning that tourists were no longer allowed to enter the country. All
hotels, lodges, and other accommodation facilities, workplaces, markets,
and places of worship were also shut down. In addition to these country-
wide restrictions, each sector (i.e., region) developed its own measures,
controls, guidelines, and procedures to guide people's daily practices. The
Uganda Wildlife Authority—the government agency mandated by law
to ensure the sustainable conservation of all wildlife within and outside
Protected Areas in Uganda—was also required to develop guidelines to
minimize contact between wildlife and humans and thus to minimize the
risk of infection in either direction. As a tour guide we interviewed noted:

> During the pandemic, the park was closed and [...] us, the guides, [...]
> were off the business and could not access any park [...] Bwindi National
> Park was closed specifically. You see, we share almost the same DNA with
> gorillas, as they possess 98.7% of humans' [DNA]. So, they had to close
> [the park] to stop the chances of spreading the disease from humans to
> gorillas. Uganda Tourism Board announced that for the meantime, all the
> national parks are going to be closed, and also the government officials
> announced it (Interviewee 7).

Gorillas (as primates) are often humanized, given their similar genetic
makeup to human beings. Indeed, some of the indistinctions between
humans and gorillas are literally embodied. Accordingly, several interviewees made references to our shared genetic makeup with gorillas and
hence to a shared vulnerability. These arguments appear to justify the
negative implications of biosecurity regimes at Bwindi, namely, the halting
of nearly all gorilla-related economic activity and related income. Border
closure and the consequent shutdown of gorilla tourism had major reper-
cussions on the livelihoods of those who were crucially dependent on this
form of tourism. During the lockdown, the Uganda Wildlife Authority's 'park revenue sharing schemes' did not generate and distribute the
usual income received from gorilla tourism in Bwindi, halting certain
conservation efforts and the associated benefits normally shared with
local communities. The lack of financial flows to the park during the

pandemic also resulted in reduced staffing and a loss of morale among the remaining park staff. The economic hardship and absence of tourism benefits arguably forced some local residents to resume poaching for survival and this was associated with the killing of the well-known gorilla 'leader' named Rafiki, seemingly out of self-defence (Uganda Wildlife Authority, 2020). The closure of the park thus had severe consequences for the human populations that had been (made) dependent on its conservation programme and the related tourism income.

During the initial lockdown, only a limited number of trackers and veterinary doctors were allowed to monitor the gorillas for any signs of ill health. These interactions were now limited to one hour as compared to the four/five hours per day usual in pre-pandemic times. Any change in gorilla health was immediately communicated to the veterinarians for further monitoring and handling. Additional adjustments were made to the preparation of the tracking and monitoring team. All staff were mandated to wear masks, and to disinfect and sanitize their footwear, and walking sticks. Additionally, the field staff were obliged to carry sanitizers and two pairs of N95 masks with them at all times. Interviews with rangers and trackers further indicated that during the pandemic, park staff were not allowed to spend time in the villages, a measure designed to prevent them from contracting the virus. In cases where it was inevitable for these staff members to go home, they would spend 14 days in quarantine before being allowed to resume their patrols in the park.

A related implication of the lockdown was that the gorillas living in the park began to get used to tourists' absence. For the park staff, this was an entirely new situation, as noted by several of the rangers, who were still required to track the gorillas in order to maintain some form of human presence, in line with established habituation goals:

> "What I came to realize was that there was some kind of behavioural change though I cannot describe it. Because these gorillas were used to seeing whites [referring to white tourists] but this time, they would only see us, and I think they were also wondering..." (Interviewee 9).

Thus, in the absence of groups of (white) tourists, the pandemic set off a process of 're-habituation' on the part of the park staff and the gorillas. Exposure to white people was notably absent and arguably impacted the gorillas' behaviour. The maintenance of human presence among the gorillas proved to be a balancing act that upheld both human-oriented and animal-oriented objectives. Specifically, it kept the gorillas safe from excessive exposure to potentially infectious human bodies while maintaining a limited embodied presence to prevent them from 're-wilding'. Other rangers referred to a rise in birth rate among gorillas during the lockdown of Bwindi:

> "There is something that happened, but I think it was coincidence, because very many [gorilla] births were recorded. The gorillas became a bit shy because they were not visited like before, [I] thought they looked healthier" (Interviewee 6).

> "We saw very many births in mountain gorillas and even also in terms of walking, gorillas did not walk much where you would find them" (Interviewee 8).

> "We did not get any gorilla that got infected; actually, we recorded a number of births during that time. Some staff whose body temperatures were detected high had to be isolated and [...] put in quarantine and no one died of Covid" (Interviewee 9).

In September 2020, six months after the initial lockdown of Bwindi, the Ugandan government opted to relax certain restrictions. Some forms of tourism mobility were resumed, albeit with strict observance of standard operating procedures. The new procedures introduced new assemblages of materialities for the tourists, such as face masks, detergents, hand sanitizers, COVID-19 test kits, and negative COVID-19 test results or certificates, among others. The tourists had to present their certificates along with their passports, their detailed travel itineraries, and proof of payment for accommodation and tourist activities such as national park entrance fees and permits. Tour operator vehicles were only cleared to carry tourists if they were licensed by the Ministry of Works and Transport, and associated with tourism companies that were registered and licensed by the Uganda Tourism Board. These vehicles also had to carry sanitizers, face masks, and first aid kits. All these procedures determined who could travel and who could not, as was the case across most tourist destinations during those months. Tourists were permitted to visit all Ugandan National Parks but not the primates' habitats, including the gorilla habitats in Bwindi. Gorilla viewing only resumed in Bwindi in

November 2021, albeit again with the enforcement of multiple restrictions. The Uganda Wildlife Authority standard operating procedures required all tourists, rangers, and guides visiting primates to wear N95 masks, surgical masks, or cloth masks with filters, while the distance from which gorillas could be viewed was increased from 7 to 10 metres. The pre-gorilla tracking activities were also adjusted to ensure the safety of both humans and animals. The interviewees reported that hand-washing and sanitizing equipment was installed at the park entrance and that all those entering the premises had to wash their hands before being allowed to access the park. The number of tourists that could be briefed at a time was reduced to eight persons maximum and during these briefings attendees were required to remain two metres apart. At the end of each briefing session, all tourists, guides, and porters were required to step through a footwear disinfecting unit before entering the forest. Furthermore, UWA staff were equipped with non-contact infrared thermometers to screen all visitors and staff for body temperature. Interviews with the rangers revealed that some tourists expressed discomfort with keeping their masks on, or refused to wear masks entirely, a source of considerable nuisance to the staff who would then have no choice but to exclude these individuals from participation:

> "[We] had visitors who never wanted to put on masks, but we would send them back [....] without seeing the gorillas and their money could not be refunded ...Viewing gorillas at 10 metres' distance, some clients complained that they were not seeing gorillas very well, cannot take selfie. But of course, the rule would not allow them to go in less than 10 metres and sometimes they would become angry" (Interviewee 5).

Despite these difficulties, park staff were in general agreement that the measures put in place to contain the spread of COVID-19 should be maintained in the future, also with a view to preventing the spread of other zoonotic diseases such as Ebola. The protection of gorilla lives— an ethic of care that was generally accepted by the park authorities and rangers, conservationists, tourism agencies, and others—was variously explained by interviewees as customary and a necessity for future generations. Seemingly, the only subjects not in line with the behavioural adaptations implied by such ethics were less compliant tourists flagged as 'stubborn clients' and 'curious young gorillas' who approached groups of visitors at a close range.

CONCLUSION

This chapter has offered an account of 'habituation' as a manifestation of a more-than-human (and potentially) affirmative biopolitics that aims to foster the lives of endangered animal bodies by making them accustomed to the presence of those who are at once their protectors and their antagonists, namely human beings. More specifically, we have argued that the bodily attunement that develops between animals and humans (including tourists) as a result of habituation is contingent upon specific spatial and temporal conditions *and* conditioning, all factors that were disrupted by the onset of the pandemic.

Our analysis of tourism-related gorilla habituation in Bwindi suggests that the lockdown of the National Park was primarily enforced to create a (semi) enclavic space and to exert biopolitical control over the human bodies entering that space, with a view to protecting the gorillas from infection and to fostering their lives in isolation. Yet, while the gorillas appear to have flourished under these new conditions—as reflected in increased birth rates—their absence as 'labour animals' from the conservation economies of tourism and animal science caused additional deprivation among human groups that had traditionally been (precariously) dependent on gorilla tourism income. The communities surrounding Bwindi that had historically been supported via tourism revenue schemes, along with park rangers and tour guides, saw a key source of income evaporate, and consequently faced hardship.

Ampumuza and Driessen (2021) have argued elsewhere that gorillas have some agency within habituation processes and that habituation can also occur naturally between gorillas and humans residing in or near the park. Our analysis here has shown that the pandemic-related absence of tourist bodies (white tourists but also others) from Bwindi caused a need for subsequent 're-habituation', because the gorillas seemingly no longer naturalized the presence of tourists in the park, unlike the presence of park rangers. This finding raises questions about the meaningfulness that tourists contribute to human–animal relations in the context of gorilla tourism, and the extent to which habituation is merely implemented to facilitate the presence of non-residential bodies in the park. If some form of 'natural' habituation previously occurred between gorillas and the indigenous communities in Bwindi, the pandemic has shown—we would contend—that an ongoing effort will always be required to rehabituate the gorillas to the presence of tourists, given the ephemeral nature of tourism visits.

Overall, the pandemic has exacerbated existing inequities that had historically been sustained by the practical implementation of conservation biopolitics and the capitalization of living matter (i.e., gorilla lives): protecting one population from infection and death was again attained at the expense of undermining the livelihoods of others. We have also highlighted in this chapter how certain human populations (e.g., the Batwa) and gorilla populations in Bwindi have historically worked through negative instances to create affirmative alternatives. Since the onset of the pandemic, the biosecurity measures have mainly targeted the circulation of human populations, including tourists, while the protection of gorilla lives has seemingly been prioritized. However, this was evidently done with a view to attaining once again the anthropocentric goal of a post-pandemic future of tourism-as-usual.

BIBLIOGRAPHY

Ampumuza, C. (2022). Living with Gorillas? Lessons from Batwa-Gorillas' convivial relations at Bwindi Forest. *Conservation & Society, 20*(2), 69–78. https://www.jstor.org/stable/27143330

Ampumuza, C., & Driessen, C. (2021). Gorilla habituation and the role of animal agency in conservation and tourism development at Bwindi, South Western Uganda. *Environment and Planning e: Nature and Space, 4*(4), 1601–1621. https://doi.org/10.1177/2514848620966502

Agamben, G. (1998). *Homo sacer: Sovereign power and bare life,* Daniel Heller-Roazen (trans.) Stanford University Press.

Agamben, G. (2004). *The open: Man and animal.* Stanford University Press.

Barua, M. (2017). Nonhuman labour, encounter value, spectacular accumulation: The geographies of a lively commodity. *Transactions of the Institute of British Geographers, 42*(2), 274–288. https://doi.org/10.1111/tran.12170

Biermann, C., & Anderson, R. M. (2017). Conservation, biopolitics, and the governance of life and death. *Geography Compass, 11*(10), e12329. https://doi.org/10.1111/gec3.12329

Blomley, T. (2003). Natural resource conflict management: The case of Bwindi Impenetrable and Mgahinga Gorilla National Parks, southwestern Uganda. In A. P. Castro AP & E. Nielsen (Eds.), *Natural resource conflict management case studies: An analysis of power, participation and protected areas. Rome: Food and agriculture organization of the United Nations,* pp. 231–250.

Bluwstein, J. (2018). From colonial fortresses to neoliberal landscapes in Northern Tanzania: A biopolitical ecology of wildlife conservation. *Journal of Political Ecology, 25*(1), 144–168. https://doi.org/10.2458/v25i1.22865

BMCT. (2023). *Batwa Empowerment.* https://bwinditrust.org/our-work/bat wa-2/

Braidotti, R. (2015). Posthuman affirmative politics. In S. E. Wilmer & A. Žukauskaitė (Eds.), *Resisting biopolitics: Philosophical, political, and performative strategies* (pp. 42–68). Routledge.

Braun, B. (2013). Power over life: Biosecurity as biopolitics. In Dobson A, Barker K and Taylor SL (Eds.), *Biosecurity: The socio-politics of invasive species and infectious diseases,* (pp. 45–57). Earthscan.

Butynski, T. M., & Kalina, J. (1993). Three new mountain national parks for Uganda. *Oryx, 27*(4), 214–224. https://doi.org/10.1017/S00306053000 2812X

Büscher, B., & Fletcher, R. (2019). Towards convivial conservation. *Conservation & Society, 17*(3), 283–296. https://www.jstor.org/stable/26677964

Chrulew, M. (2012). Animals in biopolitical theory: Between Agamben and Negri. *New Formations, 76*(76), 53–67. https://doi.org/10.3898/NEWF. 76.04.2012

Chrulew, M. (2017). Animals as biopolitical subjects. In M. Chrulew & D. J. Wadiwel (Eds.), *Foucault and animals* (pp. 222–238). Brill.

Chrulew, M., & Wadiwel, D. J. (2017). *Foucault and animals.* Brill.

Doran-Sheehy, D. M., Derby, A. M., Greer, D., & Mongo, P. (2007). Habituation of western gorillas: The process and factors that influence it. *American Journal of Primatology, 69*(12), 1354–1369. https://doi.org/10.1002/ajp. 20442

Goldsmith, M. L. (2014). Mountain gorilla tourism as a conservation tool: Have we tipped the balance? In A. E. Russon & J. Wallis (Eds.), *Primate tourism: A tool for conservation?* (pp. 177–198). Cambridge University Press.

Hardt, M., & Negri, A. (2000). *Empire.* Harvard University Press.

Hickey, J. R., Basabose, A., Gilardi, K. V., Greer, D., Nampindo, S., Robbins, M. M., & Stoinski, T. S. (2018). Gorilla beringei ssp. beringei. *The IUCN Red List of Threatened Species* 2018: e.T39999A17989719. https://doi.org/ 10.2305/IUCN.UK.2018-2.RLTS.T39999A17989719.en

Hinchliffe, S., & Bingham, N. (2008). Securing life: The emerging practices of biosecurity. *Environment and Planning A: Economy and Space, 40*(7), 1534–1551. https://doi.org/10.1068/a4054

Köndgen, S., Kühl, H., N'Goran, P. K., Walsh, P. D., Schenk, S., Ernst, N., Biek, R., Formenty, P., Mätz-Rensing, K., Schweiger, B., Junglen, S., Ellerbrok, H., Nitsche, A., Briese, T., Lipkin, W. I., Pauli, G., Boesch, C., & Leendertz, F. H. (2008). Pandemic human viruses cause decline of endangered great apes. *Current Biology, 18*(4), 260–264. https://doi.org/10.1016/j.cub.2008.01.012

Lemke, T. (2011). *Biopolitics.* NYU Press.

McAfee, K. (1999). Selling nature to save It? Biodiversity and green developmentalism. *Environment and Planning D: Society and Space, 17*(2), 133–154. https://doi.org/10.1068/d170133

Minca, C. (2006). Giorgio Agamben and the new biopolitical nomos. *Geografiska Annaler: Series B: Human Geography, 88*(4), 387–403. https://doi.org/10.1111/j.0435-3684.2006.00229.x

Mittermeier, R. A. et al. (2022, January 9). IUCN SSC Primate Specialist Group: Report 2018–2021. Report to the International Primatological Society (IPS), Quito, Ecuador.

Mokhele, I. (2022). The biopolitical production of tourism areas in Lesotho: The case of Bokong nature reserve. *Via Tourism Review, 21.* https://doi.org/10.4000/viatourism.8528

Negri, A., & Hardt, M. (2004). *Multitude: War and democracy in the age of empire.* Penguin Random House.

Ochieng, A., Koh, N. S., & Koot, S. (2023). Compatible with conviviality? Exploring African ecotourism and sport hunting for transformative conservation. *Conservation & Society, 21*(1), 38–47. https://www.jstor.org/stable/27206658

Plumptre, A. J., McNeilage, A., Hall, A. S., & Williamson, E. A. (2003). The current status of gorillas and threats to their existence at the beginning of a new millennium. In Taylor A, Goldsmith M (Eds.), *Gorilla Biology: A multidisciplinary perspective. Cambridge studies in biological and evolutionary anthropology, 35,* (pp. 141–431). Cambridge University Press.

Robbins, M. M., Gray, M., Fawcett, K. A., Nutter, F. B., Uwingeli, P., Mburanumwe, I., et al. (2011). Extreme conservation leads to recovery of the virunga mountain gorillas. *PLoS ONE. 6* e19788. https://doi.org/10.1371/journal.pone.0019788

Russon, A. E., & Susilo, A. (2014). Orangutan tourism and conservation: 35 years' experience. In A. E. Russon & J. Wallis (Eds.), *Primate tourism: A tool for conservation* (pp. 76–97). Cambridge University Press.

Samuni, L., Mundry, R., Terkel, J., Zuberbühler, K., & Hobaiter, C. (2014). Socially learned habituation to human observers in wild chimpanzees. *Animal Cognition, 17*(4), 997–1005. https://doi.org/10.1007/s10071-014-0731-6

Sandbrook, C., & Semple, S. (2006). The rules and the reality of mountain gorilla Gorilla beringei beringei tracking: How close do tourists get? *Oryx, 40*(4), 428–433.

Srinivasan, K. (2017). Conservation biopolitics and the sustainability episteme. *Environment and Planning A: Economy and Space, 49*(7), 1458–1476. https://doi.org/10.1177/0308518X17704198

Strum, S., & Nightingale, D. (2014). Baboon ecotourism in the larger context. In A. Russon & J. Wallis (Eds.), *Primate tourism: A tool for conservation?* (pp. 155–176). Cambridge University Press. https://doi.org/10.1017/CBO 9781139087407.012

Tumushabe, G., & Musiime, E. (2006). *Living on the margins of life. The plight of the Batwa communities of South Western Uganda.* ACODE Policy series, No. 17.

Uganda Wildlife Authority. (2020, June 12). *We have arrested four people over the death of Rafiki, the Silverback of Nkuringo Gorilla group in Bwindi Impenetrable National Park. They will be prosecuted in the courts of law. See statement below* [Image attached] [Tweet]. Twitter. https://twitter.com/ugwildlife/sta tus/1271351504730042368/photo/2

Uganda Wildlife Authority. (2018). *Gorilla tourism report between 2005–2018.* Uganda Wildlife Authority (UWA).

UNESCO. (2023). Bwindi Impenetrable National Park. https://whc.unesco. org/en/list/682/

van der Duim, R., Ampumuza, C., & Ahebwa, W. M. (2014). Gorilla tourism in Bwindi Impenetrable National Park, Uganda: An actor-network perspective. *Society & Natural Resources, 27*(6), 588–601. https://doi.org/10.1080/089 41920.2014.901459

Wadiwel, D. J. (2018). Biopolitics. In L. Gruen (Ed.), *Critical terms for animal studies* (pp. 79–98). The University of Chicago Press.

Weber, A., Kalema-Zikusoka, G., & Stevens, N. J. (2020). Lack of rule-adherence during mountain gorilla tourism encounters in Bwindi Impenetrable National Park, Uganda, Places Gorillas at risk from human disease. *Frontiers in Public Health, 8.* https://doi.org/10.3389/fpubh.2020.00001

Wright, P. C., Andriamihaja, B., King, S. J., Guerriero, J., Hubbard, J., Russon, A. E., & Wallis, J. (2014). Lemurs and tourism in Ranomafana National Park, Madagascar: Economic boom and other consequences. In A. E. Russon & J. Wallis (Eds.), *Primate tourism: A tool for conservation* (pp. 123–146). Cambridge University Press.

CHAPTER 5

Affirmative Alternatives to the Biopolitics of Air Travel: Actions by the Taoyuan Flight Attendant Union During the COVID-19 Pandemic

Chih-Chen Trista Lin⊚

Abstract The field of tourism offers a key to understanding past, present, and future biopolitical action implicated in the resistance against forms of biopower. Yet, the potential for tourism geographies to engage with and inform debates on biopolitics remains to be realized. To reflect that potential, this chapter examines a situated perspective surrounding tourism labour, namely the response of the Taoyuan Flight Attendant Union (TFAU) to the impact of the COVID-19 pandemic on workers in the aviation and hospitality sectors in Taiwan. I analyse the three dimensions of the union's campaigns at the exceptional time of the pandemic as "affirmative alternatives" to the biopolitical production of, and control over, air travel, health, and labour. The analytics proposed in the chapter

C.-C. T. Lin (✉)
Cultural Geography Group, Wageningen University and Research, Wageningen, Netherlands
e-mail: chihchentrista.lin@wur.nl

© The Author(s), under exclusive license to Springer Nature Switzerland AG 2023
M. Roelofsen and C. Minca (eds.), *Tourism and Biopolitics in Pandemic Times*, https://doi.org/10.1007/978-3-031-46399-0_5

open a space for understanding the "nitty-gritty" of workers' struggles to *live well*, to be governed *better*, and to establish a qualitatively different *living-with* vis-à-vis the general public. My discussion of TFAU's experiences indicates clear potential and an imperative for tourism and hospitality to nurture a sociality beyond consumerism and sexism—as an integral part of broader biopolitical and democratic struggles against governmental control and corporate capitalism.

Keywords Biopolitics · Affirmation · Labour union · Tourism and hospitality labour · Femininity · Flight attendant

INTRODUCTION

International tourism and the related air travel constitute a vast social and economic field. Since the onset of the COVID-19 pandemic, this field has served as both an actual and an imagined area of contact between foreign territories and homelands, hosts and guests, the viral and the human, and pleasure and tension. Air travel in particular has emerged as a prominent political arena, where we have witnessed the general intensification of (bio)security and border surveillance apparatuses by many national governments, and where multiple biopolitical caesurae were enacted in deciding which subjects were worthy of health and mobility and which were not (Adey et al., 2021; Rose-Redwood et al., 2020). It is in this sense that issues pertaining respectively to vital geographies and biopolitical geopolitics—two key foci of biopolitical scholarship within geography, as suggested by Rutherford and Rutherford (2013)—have jointly come to the fore within academic debates.

Crucially, tourism has also long been the quintessential sector where hyperprofits are generated on the back of immaterial, affective, and emotional labour (Fletcher et al., 2023; Minca, 2010; Veijola, 2009; see also Ioannides & Zampoukos, 2018). Michael Hardt and Antonio Negri (2000, 2004, 2009) famously stated that, in the new globalized economy, affective labour should be viewed as the "locus of resistance" against all forms of control over life (Žukauskaitė & Wilmer, 2016, p. 7). Hence, the field of tourism offers a key to understanding past, present, and future biopolitical action implicated in the resistance against all forms of biopower. Yet, the potential for studies of tourism labour to engage with (or inform) broader debates on biopolitics in this regard remains to be more fully realized (cf. Fletcher et al., 2023). For example, Veijola and Jokinen observed that, in the globalized economy, "new work" involving tourism labour both requires and develops skills such as anticipation, coordination, and "the ability to venture" (2008, p. 177). Might this specific quality of tourism and hospitality labour contribute to new forms of social and political life—in opposition to the present coercion, exploitation, and control of populations that many have rightly denounced (see e.g., Adey et al., 2021; Aradau & Tazzioli, 2020; Baum et al., 2020; Roelofsen & Minca, 2018)?

Speaking to these major areas of concern, which draw together tourism and biopolitics, this chapter examines a situated perspective surrounding collectivized labour in tourism and air travel. I focus on the response of the Taoyuan Flight Attendant Union (TFAU) in Taiwan to the impact of the COVID-19 pandemic on workers in the aviation and hospitality sectors. As the primary means of cross-border transportation, air travel in Taiwan is both commoditized and highly commercialized, while remaining an essential public utility. TFAU first called on airlines and the Taiwanese government to issue appropriate health, safety, and prevention protocols for cabin crew in early January 2020. My discussion in the chapter is based on discourse analysis of all TFAU's subsequent press releases, social media posts, audiovisual campaign materials, documents, and other related website content from 2020 to 2022, as well as on an interview with two official representatives of the union. I also review governmental press releases and other public statements on air travel and airline labour during the same period, along with salient news stories covered by the digital outlets of over a dozen major Taiwanese news media organizations.

As I detail below, the union's campaigns during the pandemic may be seen as having three distinct dimensions, from (i) urging the public authorities to take over the governance of disease prevention, to (ii) making detailed policy recommendations concerning the management of labour during the pandemic, to (iii) addressing issues of stigmatization and discrimination. Over the following sections of the chapter, I analyse these three dimensions as "affirmative alternatives" to the biopolitical production of, and control over, air travel, health, and labour in Taiwan at the exceptional time of the pandemic. In light of the "biopolitical turn" in the humanities and the social sciences over the past two decades or so, scholars have questioned the analytical value of theories of biopolitics when they come to act as "all-encompassing visions of emergency and violence" (Anderson et al., 2020, p. 629). More specifically in relation to geographers' burgeoning engagement with biopolitics, some authors have cautioned against the tendency to advance reductive biopolitical readings that link "all power to the dark machinations of necro-politics" (Barnett, 2012, p. 380). In response, there have been calls to direct inquiry towards "affirmative biopolitics" or "affirmative alternatives" to biopolitical control, both within and beyond geography (Hannah, 2011; Rutherford & Rutherford, 2013; Lin et al., 2018; see also Prozorov, 2019; Yetiskin, 2022). These calls for "affirmation" are invariably underpinned by a focus on, and some new elaboration of, the affirmative and productive nature of biopower[1]—understood as a political power that fosters, regulates, and regularizes the life of the population under its governance, such that this life unfolds in specific ways (see, among others, Berlant, 2011; Protevi, 2009a, 2009b). In this context, "affirmative alternatives" refer to the politics of active engagement with and appropriation of this facet of biopower by living beings, instead of the passive submission to or acceptance of biopower that is generally presumed and often real (Lemke, 2011, p. 120). As an analytical enterprise, focusing on affirmation offers the potential to open a new path, as envisioned by Thomas Lemke (2011, p. 122), that "leads beyond the futile choice between the trivialization and the dramatization of biopolitical phenomena".

My own analytical focus on "affirmative alternatives" draws inspiration from various strands of feminist approaches to biopolitics. Whereas Hardt and Negri's conceptualizations of affective labour—as capable of affirmatively appropriating the power of life—have become popular in recent years, many have critiqued their discussion as inspirational yet vague (see for example, Chari, 2017; Žukauskaitė & Wilmer, 2016). Feminist scholars, among others, have noted that Hardt and Negri's work does not fully address the complexity of contemporary forms of biopolitical production and control, nor of affirmative political struggles to resist them (Berlant, 2011; Braidotti, 2013). In particular, Rosi Braidotti has highlighted the need for "context-specific, historicized accounts" of affirmative politics of life (2016, p. 32; see also Lemke, 2011, p. 122). On the one hand, such accounts should detail the production of and control over

---

[1] Encapsulated within Michel Foucault's most famous statement about biopower is its characterization as a power to "make live" ([1976]2003, p. 248), which co-exists with the power of the sovereign to take life and "make die".

life amidst shifting (geo)political, economic, socio-cultural, and techno-logical divides among living beings; on the other hand, special attention must be paid to biopolitical actions as "transformative experimentations with new arts of existence and ethical relations" above and beyond human desire for/as *bios* (Braidotti, 2016, p. 30).

Another key effort to theorize affirmation has been offered by Kathi Weeks (2007, 2011), who expanded Hardt and Negri's discussion of the post-Fordist biopolitical organization of affective labour and life. Indeed, Weeks advances a well-articulated and helpful discussion of how the potential becoming of the affective labourer "self" can serve as an immanent critique of and resistance to biopower. Explicitly evoking generations of feminist critiques of traditional patriarchal family values, the gendered division of labour, and the distinction between the public and private spheres, Weeks formulates a clear strategy for this "self". She suggests that it is neither necessary nor desirable to retreat into a roman-ticized authentic "self" or into the often-idealized private domain (e.g., family; home) which fall outside of historical and ongoing biopolitical control over life and affective labour. Rather, Weeks points up the need to "struggle for a different quality of experience", precisely from within our current biopolitical condition (2007, p. 247). More specifically, this strategy encourages the affirmation of "a self that one might wish to become"—a self that is firmly rooted in "the self at work" with its current level of conditioning (2007, p. 248):

> Affirmation in this sense requires that we not refuse what we have now become after measuring ourselves against the standard of what we once were or what we wish we had become, but affirm what we are and will it, because it is also the constitutive basis from which we can struggle to become otherwise (Weeks, 2011, p. 201).

TFAU's Affirmative Alternatives to Pandemic Biopolitics

Contesting Biopolitical Caesurae

TFAU has been in operation since 2015.[2] At the beginning of the pandemic, its first public actions were driven by the lack—at that time—of a proactive response, in terms of in-flight infection prevention, on the

---

[2] For further background on TFAU's activities, see The Reporter (2016).

part of Taiwan's two main airlines' (China Airlines and Eva Air). The union's initial campaigns in early 2020 were thus aimed at directing the attention of the government and of the broader public to the need for in-flight hygiene and disease prevention for cabin crew and air passengers via the use of face masks, gloves, protective glasses/goggles, and so on. Taiwan's Central Epidemic Command Center (CECC) was officially set up to address the novel coronavirus on 20 January 2020, marking the emergence of the biomedical regime for governing COVID-19. TFAU took immediate action to assert its oversight over flight crews and airlines, via a first attempt to contact CECC on 21 January 2020. However, TFAU's occupational health representative, Yu-Jie Gon, recalled in an interview with me that the centre did not initially pay enough attention to in-flight disease prevention and lacked the knowledge to manage related aspects of cabin crews' working conditions, such as face-to-face contact with passengers during the distribution of meals and other services.

For several months following the onset of the pandemic, TFAU did not feel that the occupational health, labour rights, and other related needs and interests of cabin crews were being sufficiently taken into account in the drawing up of the government's quarantine measures and vaccination policy (see for example The Reporter, 2020). The union's actions at this stage included attempts to engage with the executive and regulative biopower of government agencies (such as the CECC), in addition to its contact with the usual authority with responsibility for civil aviation (Civil Aeronautics Administration, or CAA, under the Ministry of Transportation). Specifically, TFAU urged and demanded that the collective labour and health of cabin crews should be duly valued in light of the pandemic, and managed accordingly. In addition, by frequently issuing press releases and circulating statements through social media and other channels, TFAU sought to (re)subjectivize cabin crews as more than glamourized, feminized hospitality workers, a frequent perception among the general public (see the next sections). Indeed, TFAU members repeatedly represented themselves as subjects who were committed to protecting and defending Taiwan—in terms of the health and safety of air passengers, and by extension of all the country's residents and citizens—from potential breaches in national (bio)security. They might thus be seen as evoking the sentiment and logic of "defending the society" (Foucault, [1976]2003, p. 61) via this new positioning of themselves as subjects: as indispensable frontline workers, they were not only in need, but also deserving of resources and assistance to prevent disease. The union's campaigns contributed to the CECC's decision towards the end of March 2020 that all flight crews should be supplied with and required to use full protective gear.

Intervening in Actuarial, Technocratic, and Corporate Management

From the second half of 2020 through 2022, TFAU remained active in terms of responding to the constantly shifting regulation of cabin crews' working and quarantine conditions, in combination with governmental and corporate disciplining of flight attendants' gendered labouring bodies over this period. In relation to the new biomedical governance regime that quickly became established in Taiwan, these TFAU actions were somewhat different in nature from the earlier initiatives described above. For instance, TFAU regularly joined forces with the Taoyuan Union of Pilots (TUP) and the two airlines' in-company unions to contest the government's decisions concerning testing, personal protective equipment, vaccination, and quarantining, with varying degrees of success. One particular campaign targeted the CECC's and CAA's strict quarantining of flight crews' bodies while both on and off duty, whether inside or outside the country. As part of this campaign, TFAU and the other unions emphasized the high number of days that flight crews spent in isolation at home and abroad (in some cases over 200 days a year). They drew on scientific research to posit that isolation placed a strain on mental health, and that mental health issues among cabin crew members put flight safety at risk, whereas only a low rate of infected cases had been confirmed among this group of workers (Control Yuan Taiwan, 2022). Implicit in these arguments was a critique of the actuarial practices[3] and pathogenic (as opposed to "salutogenic") logic behind the rules imposed by the authorities.

In the face of rising technocracy in Taiwan over this period (Chen, 2020; Markarychev & Wishnick, 2022), the union both participated in and disrupted the government's use of the media and digital information technology. For example, TFAU representatives noted that they learned to work with CECC's information dissemination techniques and technologies. As the CECC held daily and well-publicized (live-streamed and televised) press briefings about public health control, the union got its

---

[3] For further discussion of this topic, see for example Berlant (2011).

demands included in the discussion by releasing statements to the news media, which journalists could then raise with the health authority at the briefings. Another example involved the "Medi-Cloud system"—part of the administration of Taiwan's national health insurance—where the data of individuals seeking medical care is stored. During the pandemic, the CECC arranged for this system to be connected with other databases, such as those of the National Immigration Agency, thereby creating a special system for tracking an individual's occupation, travel, and contact history (CDC Taiwan, 2022; Makarychev & Wishnick, 2022). Owing to their occupation, cabin crews found themselves marked in this system as a "high risk group" alongside medical and health care workers. Concerned about possible breaches of data confidentiality, the union contested this particular use of digital technologies. As the pandemic evolved over 2022, TFAU was able to pressure government authorities into finally eliminating this tracking of "high-risk" workers.

All the while, TFAU sought to elicit changes *to* and the (re)constitution *of* the corporate management of employees' bodies and affectivity during the pandemic (on corporate management see Chang, 2016; Hochschild, 2012; Lin, 2015; see also Cheng et al., 2022). Especially in the early months of 2020, the two major airlines in Taiwan were criticized for hesitating to adapt their longstanding rules requiring their flight attendants to be immaculately groomed. The management was concerned that wearing personal protective gear against the novel coronavirus would obscure the important, idealized feminine facial appearance of flight attendants.[4] In an interview conducted for the purposes of this research, TFAU representative Gon alluded to her airline initially refusing any proposal for face coverings (e.g., protective eyewear), "for fear of affecting the aesthetics". At that time, TFAU strategized to convince the company by collecting evidence that foreign airlines were permitting the use of protective gear among their cabin crews—even providing photographs illustrating changes in the appearance of these other flight attendants. Gon recalled: "Our argument to the company was that these were foreign airlines who were listed higher than us in prestigious international airline rankings. Clearly, these airlines were not worried that the change in grooming and aesthetics would affect the quality of work. Looking pretty was not the only thing that should matter".

---

[4] The vast majority (92%) of flight attendants with Taiwanese airlines are women, according to the CAA's records.

Later, when the CECC officially ordered the use of full protective equipment among cabin crew, as mentioned above, TFAU noted the airlines' new demands for attaining "affective capture". Flight attendants were asked to take special care with grooming their hair, to apply eye makeup and to "smile at the customers with their eyes". During our interview, Gon stated that: "In-flight service managers would stress the importance of hair and eye makeup during the cabin crew pre-flight briefings. Even with the face mask and everything, they still wanted you to be pleasing to the eye". While complaining about an exceptional, unacceptable case of a manager who even demanded and inspected the wearing of lipstick under face masks, Gon was in principle "fine" with these new demands for affective labour. Admitting that, in her opinion, makeup was part of what "being professional" and "dedication" were about in her occupational field, Gon added another positive note: "I do think that grooming these components [(eyes and hair)] of your appearance makes you feel more 'energized' at work".

From Living-well to Living-with

My discussion up to this point has portrayed the "complex strategic situation" (Braidotti, 2016, p. 32) within which TFAU mobilized a relatively unique form of hospitality and transportation labour with a view to enhancing work, life, and health in terms of *bios*—mainly during the pandemic but also in the context of broader governmental and corporate control over worker-residents in Taiwan (Pan, 2020; The Reporter, 2016). From TFAU's perspective, its active involvement in the formation of Taiwan's biosecurity and occupational health apparatuses, its appropriation of subjectivities and technologies, and the related, seemingly "interminable[,] political negotiations" (Cheah, 2007, p. 112), all played a crucial part in bringing about meaningful and desirable alternatives to the forms of biopolitical control being experienced by its members. Thus, TFAU's affirmative biopolitical actions reveal an "immanent standpoint of critique" which is "at once fully implicated in, but nonetheless potentially set against the spaces, relations, and temporalities [...] dominated by work" (Weeks, 2007, p. 247).

Weeks (2007) has envisioned such actions and their logic of immanence as key to a "liberatory" project that departs from post-Fordist biopolitical regimes of work and life. This project, Weeks proposes, generates affective relations "of equality and autonomy rather than hierarchy and command" (2007, p. 247). Relatedly, Braidotti has argued that such a liberatory project is possible only when it goes beyond merely mobilizing people around the ideal of *living-well* and in the name of *bios*. The functioning and contestations of biopower in practice (re)shape *bios*, separating it from *zoe*,[5] citizens from other citizens, citizens from non-citizens. Among late capitalism's "ever-shifting waves of genderisation and sexualisation, racialisation and naturalisation of multiple 'others'" (Braidotti, 2009, p. 104), it is essential that the liberatory project should (re)constitute a widespread, ethical mode of "*living-with-others*", in terms of revisiting and overcoming the separation between *bios* and *zoe*, between "self" and "other", and between "human" and "non-human" lifeforms. This would involve ethical practices with the power to transform social interaction among fellow living beings beyond the proliferation of discrimination, exclusion, and exploitation that marks late-capitalist societies (see also Righi, 2011). It is in relation to this "*living-with-others*" that I wish to analyse TFAU's campaigns addressing the stigmatization of, and discrimination against, flight crews as "potentially dangerous vectors of COVID-19" (Iaquinto, 2020, p. 176).

Since the popularization of air travel among the Taiwanese in the 1980s and the 1990s, flight attendants in Taiwan have been strongly associated with idealized femininity, as well as with a glamourized working life presumed to be full of opportunities for travel and foreign luxuries (Chang, 2008). Despite being perceived by the general public as relatively socio-economically privileged, TFAU members have been working to expose the little-known exploitation and disciplining that they face in their profession. They have drawn public attention to the control that is exerted over their labour and affectivity in order to profit the airlines (a pattern that is also reflected in the earlier part of this discussion), as well as the rampant sexism in their work environment (see also Chang, 2016). Increasingly, the union has also become vocal about working conditions in other occupations in Taiwan, thereby supporting various other ongoing unionization projects.

Nevertheless, during the pandemic, TFAU was faced with a new and additional challenge when (the few) cases of flight crews who contracted COVID-19 were sensationalized, leading to popular outrage in Taiwan. As misconceptions circulated about crews being privilege-seeking and

---

[5] Pre- and non-human life (Braidotti, 2009).

frivolous in their approach to physical distancing, quarantine, vaccination, and other disease prevention protocols, related innuendos and rumours reproduced the misogynistic and sexist objectification of female crews. When I interviewed TFAU's representatives in April 2022, they reflected at length on the urgent need they had identified at that time for a series of campaigns against misinformation, stigmatization, and discrimination. During these campaigns, TFAU ensured that its members and allies appeared in media broadcasts that addressed these issues, and that audiovisual campaign materials were disseminated via a diverse range of media channels. Here, the union relied on its members' sharing of their actual personal and professional experiences to convey to the public the difficulties being faced, while emphasizing the crews' professionalism and commitment to "protecting the homeland" against COVID-19.

These materials combined a range of narratives and evoked various forms of affect—from crew members' passion and enthusiasm for their job to balancing the demands of work with caring for family, to a mutual interest in positive and respectful encounters between flight crews and potential future air passengers. TFAU thereby appealed to the public's sense of caring and compassion, seeking recognition and empathy for cabin crews (see Fig. 5.1). Even as union members were tasked with emotional and affective labour—although this was not a complete departure from the normal requirements of their work—, the campaign allowed them to experiment with their collective capacity to relate to the general public in affirmative yet potentially transformative ways: in other words, the production of multiple narratives and kinds of affect based on the members' lived experiences might be seen as an attempt to establish a qualitatively different *living-with* vis-à-vis the general public. Specifically, the union was attempting to modify the views of a public that has long adopted a sexist and consumerist *othering* gaze towards cabin crews. While further research is required to determine the effects of the campaign among the residents and citizens of Taiwan, its content may be seen as facilitating "multiple forms of belonging and complex allegiances" (Braidotti, 2016, p. 51) between airline labourers and the general public, potentially transcending the sphere that is conditioned and contained by biopower or corporate interests as discussed above.

Fig. 5.1 A TFAU member appealing to the public in a campaign video produced by the Unions (*Source* TFAU)

CONCLUDING REMARKS

At the outset of the pandemic, some scholars—including de Kloet et al. (2020)—lamented what they saw as a general tendency on the part of local populations to immediately comply with the biopolitical control measures implemented by the authorities in China, Hong Kong, and Taiwan. However, I am wary of such broad-brush, hasty readings. Biopolitical analyses cannot afford to obscure the heterogeneity and nuances that characterize the actions and actors implicated in biopolitical control. Crucially, biopolitical analyses need to shed light on the very exercise and composition of biopower, precisely as it plays out through lived relations in situ. Accordingly, this chapter has shown that affirmation is a potentially fruitful analytical lens that may be brought to bear upon airline workers' lived, ethical appropriation of the biopolitical regularization of their lives. I have discussed the details of this appropriation, as described by union representatives, as it concerns workers' experimentation with subjectivization and technologies in response to the government's biopolitical caesurae and its actuarial and technocratic management of their lives. I have also examined the workers' efforts to intervene in the corporate regularization of their affectivity. Such analytics first and foremost open a space for understanding the "nitty–gritty" of tourism and hospitality workers' struggles to *live well*, and to be governed *better*. Arguably, my discussion of TFAU's contestation of government and corporate management policies lends support to Young's (2022) recent proposals concerning the complex relations between given biopolitical apparatuses, and grassroots or community biopolitical action. More often than not, action is intertwined with existing biopolitical apparatuses, if not constitutive of them.

At the same time, however, an analytical focus on affirmation should also prompt us to consider not merely *living-well* but also *living-with*. By this, I mean that actions and struggles such as those we have reviewed may represent a step towards the liberatory project of *living-with-others*— which could eventually lead to the reconstitution of a sociality that transgresses the *bios/zoe* divide (Braidotti, 2009) and the bounds of biopower. Here, I have discussed TFAU's campaigns for a "nonexploitative sociality" (Righi, 2011, p. 168) in the sphere of travel and hospitality, which paralleled the union's intervention in disease and (bio)security management. TFAU's members and the campaigns clearly reflect awareness, determination, and commitment to identifying and reworking the "addictive and enjoyment-frustrating consumerism" that is underpinned by sexist *other*ing and exploitation of workers' affectivity (Braidotti, 2016, p. 41). Ultimately, any overarching liberatory project associated with nonexploitative *living-with-others* must necessarily be a long-term project and will require more than a set of campaigns by a union. Nevertheless, TFAU's members are representative of contemporary affective labour in travel, hospitality, and tourism; as such, their perspectives and experiences indicate clear potential and perhaps even an imperative for tourism and hospitality labour to nurture its collective capacity to usher in a sociality beyond consumerism and sexism. Importantly, as illustrated in the case examined here, this should be an integral part of broader, contemporary biopolitical and democratic struggles against governmental control and corporate capitalism.

Competing Interests

The author has no conflicts of interest to declare that are relevant to the content of this chapter.

References

Adey, P., Hannam, K., Sheller, M., & Tyfield, D. (2021). Pandemic (im)mobilities. *Mobilities, 16*(1), 1–19. https://doi.org/10.1080/17450101.2021.1872871

Anderson, B., Grove, K., Rickards, L., & Kearnes, M. (2020). Slow emergencies: Temporality and the racialized biopolitics of emergency governance. *Progress in Human Geography*, *44*(4), 621–639. https://doi.org/10.1177/030913 2519849263

Aradau, C., & Tazzioli, M. (2020). Biopolitics multiple: Migration, extraction, subtraction. *Millennium: Journal of International Studies*, *48*(2), 198–220. https://doi.org/10.1177/0305829819889139

Barnett, C. (2012). Geography and ethics: Placing life in the space of reasons. *Progress in Human Geography*, *36*(3), 379–388. https://doi.org/10.1177/0309132510370672

Baum, T., Mooney, S. K. K., Robinson, R. N. S., & Solnet, D. (2020). COVID-19's impact on the hospitality workforce—new crisis or amplification of the norm? *International Journal of Contemporary Hospitality Management*, *32*(9), 2813–2829.

Berlant, L. (2011). *Cruel optimism*. Duke University Press. https://doi.org/10.1108/IJCHM-04-2020-0314

Braidotti, R. (2006). Affirmation versus vulnerability: On contemporary ethical debates. *Symposium: Canadian Journal of Continental Philosophy*, *10*(1), 235–254. https://doi.org/10.5840/symposium200610117

Braidotti, R. (2009). Locating Deleuze's eco-philosophy between Bio/Zoe-power and necro-politics. In R. Braidotti & C. Colebrook PH (Eds.), *Deleuze and law* (pp. 96–116). Palgrave Macmillan.

Braidotti, R. (2013). *The posthuman*. Polity Press.

Braidotti, R. (2016). Posthuman affirmative politics. In SE., Wilmer A. Žukauskaitė (Eds.), *Resisting biopolitics: Philosophical, political, and performative strategies* (pp. 30–56). Routledge.

CDC Taiwan (2022). *TOCC information recorded on NHI Medi-Cloud system adjusted in response to coronavirus situation*. Taiwan Centers for Disease Control. Retrieved June 28, 2023, from https://www.cdc.gov.tw/En/Bulletin/Detail/iia63fOoDRQnW9VUkBzlag?typeid=158

Chang, C.-Y. (2008). *Love, struggle, and fly: The lived experiences of Taiwanese female flight attendants*. MA thesis. University of Florida.

Chang, C.-Y. (2016, June 29). *Gender, emotional labour and the challenges faced by flight attendants in Taiwan*. Initium Media. Retrieved June 28, 2023, from https://theinitium.com/article/20160629-opinion-chinaairlines2/

Chari, S. (2017). The blues and the damned: (Black) life-that-survives capital and biopolitics. *Critical African Studies*, *9*(2), 152–173. https://doi.org/10.1080/21681392.2017.1331457

Cheah, P. (2007). Biopower and the new international division of reproductive labour. *Boundary 2*, *34*(1), 79–113. https://doi.org/10.1215/01903659-2006-028

Chen, J.-S. (2020). Biopolitics of a pandemic prevention community. *Taiwanese Journal of Sociology*, *67*, 237–246. https://doi.org/10.6786/TJS.202006_ (67).0008

Cheng, T.-M., Hong, C.-Y., & Zhong, Z.-F. (2022). Tourism employees' fear of COVID-19 and its effect on work outcomes. *Current Issues in Tourism*, *25*(2), 319–337. https://doi.org/10.1080/13683500.2021.1978952

Clough, P. T., & Halley, J. (Eds.). (2007). *The affective turn*. Duke University Press.

Control Yuan Taiwan. (2022). *Investigation Report no. 111社調0025*. The Control Yuan Taiwan. Retrieved June 28, 2023, from https://www.cy.gov.tw/CyBsBoxContent.aspx?n=133&s=18001

Fletcher, R., Blanco-Romero, A., Blázquez-Salom, M., et al. (2023). Pathways to post-capitalist tourism. *Tourism Geographies*, *25*(2–3), 707–728. https://doi.org/10.1080/14616688.2021.1965202

Foucault, M. ([1976]2003). *Society must be defended. Lectures at the Collège de France, 1975–76*. Macey D (Trans.). M. Bertani, & A. Fontana (Eds.). Picador.

Hannah, M. G. (2011). Biopower, life and left politics. *Antipode*, *43*(4), 1034–1055. https://doi.org/10.1111/j.1467-8330.2010.00840.x

Hardt, M., & Negri, A. (2000). *Empire*. Harvard University Press.

Hardt, M., & Negri, A. (2004). *Multitude*. Penguin Press.

Hardt, M., & Negri, A. (2009). *Commonwealth*. Harvard University Press.

Hochschild, A. R. (2012). *The managed heart*. University of California Press.

Iaquinto, B. L. (2020). Tourist as vector: Viral mobilities of COVID-19. *Dialogues in Human Geography*, *10*(2), 174–177. https://doi.org/10.1177/2043820620934250

Ioannides, D., & Zampoukos, K. (2018). Tourism's labour geographies: Bringing tourism into work and work into tourism. *Tourism Geographies*, *20*(1), 1–10. https://doi.org/10.1080/14616688.2017.1409261

de Kloet, J., Lin, J., & Chow, Y. F. (2020). 'We are doing better': Biopolitical nationalism and the COVID-19 virus in East Asia. *European Journal of Cultural Studies*, *23*(4), 635–640. https://doi.org/10.1177/1367549420928092

Lemke, T. (2011). *Biopolitics*. E.F. Trump (Trans). New York University Press.

Lin, C.-C.T., Minca, C., & Ormond, M. (2018). Affirmative biopolitics: Social and vocational education for Quechua girls in the postcolonial "affectsphere" of Cusco, Peru. *Environment and Planning d: Society and Space*, *36*(5), 885–904. https://doi.org/10.1177/0263775817753843

Lin, W. (2015). 'Cabin pressure': Designing affective atmospheres in airline travel. *Transactions*, *40*(2), 287–299. https://doi.org/10.1111/tran.12079

Markarychev, A., & Wishnick, E. (2022). Anti-pandemic policies in Estonia and Taiwan: Digital power, sovereignty and biopolitics. *Social Sciences, 11*(3), 112. https://doi.org/10.3390/socsci11030112

Minca, C. (2010). The island: Work, tourism and the biopolitical. *Tourist Studies, 9*(2), 88–108. https://doi.org/10.1177/1468797609360599

Pan, S.-W. (2020). Still trapped between the state and management: Unions and worker representation in Taiwan in an era of globalization. In B.-H. Lee, S.-H Ng, & R. Lansbury (Eds.), *Trade unions and labour movements in the Asia-Pacific region* (pp. 255–272). Routledge.

Protevi, J. (2009). The Terri Schiavo case: Biopolitics, biopower and privacy as singularity. In R. Braidotti, & C. Colebrook PH (Ed.), *Deleuze and law* (pp. 59–72). Palgrave Macmillan.

Prozorov, S. (2019). *Democratic biopolitics: Popular Sovereignty and the power of life*. Edinburgh University Press.

Righi, A. (2011). *Biopolitics and social change in Italy: From Gramsci to Pasolini to Negri*. Palgrave Macmillan.

Roelofsen, M., & Minca, C. (2018). The superhost. Biopolitics, home and community in the *Airbnb* dream-world of global hospitality. *Geoforum, 91*, 170–181. https://doi.org/10.1016/j.geoforum.2018.02.021

Rose-Redwood, R., Kitchin, R., Apostolopoulou, E., Rickards, L., Blackman, T., Crampton, J., Rossi, U., & Buckley, M. (2020). Geogrpahies of the COVID-19 pandemic. *Dialogues in Human Geography, 10*(2), 97–106. https://doi.org/10.1177/2043820620936050

Rutherford, S., & Rutherford, P. (2013). Geography and biopolitics. *Geography Compass, 7*(6), 423–434. https://doi.org/10.1111/gec3.12047

The Reporter. (2016, June 25). *China Airlines flight attendant strike*. The Reporter. Retrieved June 28, 2023, from https://www.twreporter.org/a/china-airlines-flight-attendants-1

The Reporter. (2020, March 31). *Flight attendants, care workers, delivery workers as high-risk groups: Qualification of COVID-19 as a work-related injury*. The Reporter. Retrieved June 28, 2023, from https://www.twreporter.org/a/covid-19-infectious-disease-occupational-injury

Veijola, S. (2009). Gender as work in the tourism industry. *Tourist Studies, 9*(2), 109–126. https://doi.org/10.1177/1468797609360601

Veijola, S., & Jokinen, E. (2008). Towards a hostessing society? Mobile arrangements of gender and labour. *NORA—Nordic Journal of Feminist and Gender Research, 16*(3), 166–181.

Weeks, K. (2007). Life within and against work: Labour, feminist critique, and post-Fordist politics. *Ephemera, 7*(1), 233–249.

Weeks, K. (2011). *The problem with work. Feminism, Marxism, antiwork politics and postwork imaginaries*. Duke University Press.

Yetiskin, E. (2022). Biopolitics of "acquired immunity": The war discourse and feminist response-abilities in art, science, and technology during COVID-19. *OMICS A Journal of Integrative Biology, 26*(10), 552–566. https://doi.org/10.1089/omi.2022.0091

Young, I. (2022). Biopolitics, citizenship, and inequalities in HIV assemblages. *Dialogues in Human Geography, 12*(1), 124–128.

Žukauskaitė, A., & Wilmer, S. E. (2016). Introduction. In S. E. Wilmer, & A. Žukauskaitė (Eds.), *Resisting biopolitics: Philosophical, political, and performative strategies* (pp. 1–18). Routledge. https://doi.org/10.1177/20438206211054595

CHAPTER 6

Afterthoughts

Claudio Minca and *Maartje Roelofsen*

Abstract In this concluding chapter the main findings of the book are summarized. The chapter also provides suggestions for potential directions that the biopolitical in tourism may take in the near future.

Keywords Biopolitics · Tourism · COVID-19 pandemic

As noted in the opening chapter, the four brief case studies making up this collection represent a first tentative attempt to empirically engage

C. Minca
Department of History and Cultures, University of Bologna, Bologna, Italy
e-mail: claudio.minca@unibo.it

M. Roelofsen (✉)
Open University of Catalonia, Barcelona, Spain
e-mail: mroelofsen@uoc.edu; maartje.roelofsen@wur.nl

Department of Economics and Business, Universitat Oberta de Catalunya, Barcelona, Spain

Cultural Geography Group, Wageningen University and Research, Wageningen, The Netherlands

M. Roelofsen and C. Minca (eds.), *Tourism and Biopolitics in Pandemic Times*, https://doi.org/10.1007/978-3-031-46399-0_6

103

with the biopolitical in tourism, while at the same time exploring different theoretical approaches to biopolitics, in line with representative articulations of existing debates within the humanities and the social sciences: from Esposito's immunisation paradigm to a hybrid discourse and narrative analytical approach informed by Foucault's and Agamben's readings of biopolitics, and from more-than-human understandings of the biopolitical to affirmative stances inspired by the feminist work of Braidotti and Berlant. Needless to remark, much more could have been done to include further perspectives from contemporary debates on the biopolitical and to examine a more diverse range of cases from more regions and of more different types. Nevertheless, within the limited and partial scope of this editorial project, we advance a few general considerations with a view to hopefully encouraging further research in this line of inquiry.

First, the pandemic appears to have emphasised existing biopolitical features of tourist experience, especially by accentuating the diverse and selective nature of tourist mobilities. Second, the health crisis threw the role of workers in the 'making of tourism' into dramatic relief: as all the chapters in the volume have clearly shown, some workers were exposed to the risk of losing their jobs and (in some cases) concurrently to the risk of contracting the virus in the course of extensive daily contact with tourists, especially during the early stages of the pandemic. Third, Chapters 1, 2, and 4 pointed up the frequency and ease with which tourist spaces may be converted into spaces of exception: from the conversion of cruise ships into isolated sanatoria, to the Covid-free island experiment, to the more-than-human dimension of the management of a famous national park in Uganda. Fourth, and perhaps most importantly, what all the chapters seem to suggest is that, despite the measures implemented by many governments to protect the broader population from the virus and to produce putatively safe spaces for tourists, the persistence of neoliberal policies aimed at supporting the industry and the irresistible desire of the tourists to act 'as if' the epidemic were not taking place, led to the implementation of quasi-thanatopolitical strategies that inherently implied the sacrifice of the health of some-bodies to favour the mobility and the consumption of other (more privileged, but still at risk) bodies. The biopolitical analyses of the four cases discussed in this book have therefore brought into valuable focus how certain strategies implemented by governments and industry had a direct impact on the safety and the well-being of selected bodies and were part of a risk calculus concerning the mass of tourists allowed to travel.

Finally, tourism in pandemic times also leaned towards new forms of biopolitical experimentation, aimed at responding to the emergency situation but also at imagining possible new tourism futures (on 'tourism futures', see, among others, Tzanelli, 2023a, 2023b; also Brouder, 2020; Sheller, 2021): from the production of added value for enclavic and (presumedly) immune tourist spaces (Chapter 2) to new forms of resistance on the part of tourism workers (Chapter 5); from the media's focus on the safety and well-being of highly racialised and privileged populations (Chapter 3) to persistent colonial understandings of nature and the associated biopolitical relationship with the realm of the non-human (Chapter 4).

It is difficult to say at this point whether any of this experimentation will prove resilient to the point of wielding a lasting impact on the post-Covid tourism world (the central topic of a number of works, including among many others, Brouder et al., 2020; Lew et al., 2020; Sheller, 2021). What we may perhaps venture to say is that the pandemic has confirmed and further exalted the importance of tourism and tourist mobilities within the lifestyles and geographical imaginations of ever-expanding populations; while at the same time, it has exacerbated the mobility injustice produced by contemporary neoliberal economic regimes and frequently decried by Sheller and Urry (2006, 2016), including in relation to tourism (see also, Sheller, 2018).

The biopolitical aspects of tourism are possibly set to become all the more important, given that the intensive individual profiling that was implemented during the pandemic will almost certainly remain in place, and that the 'trusted travellers' of the future will be determined by biopolitical affordances that will have much in common—and already have much in common—with bordering practices and infrastructures aimed at halting and containing other mobile bodies (such as refugees or other 'people on the move'—see Dijstelbloem, 2021). State intervention in defining, monitoring, and controlling which subjects are allowed to travel as tourists and how they may travel—combined with the increasingly sophisticated ways in which the industry will select and encourage specific bodies to participate in the tourist economy—may well establish biopolitics as a crucial factor in the production of future tourist geographies and spatialities.

Indeed, the pandemic has represented an opportunity to revisit practices and traditionally dominant conceptualisations of tourism (see Cresswell, 2021; Minca & Roelofsen, 2023; Mostafanezhad et al., 2020;

Sheller, 2021, among numerous other interventions) as a form of selective mobility that is closely related to the history of colonialism and to forms of discrimination based on gender, class, race, and anthropocentric views of nature. In this sense, the recovery of the tourist industry following the pandemic does not look to be taking shape in promising ways, given the re-emergence of overtourism across the globe and the re-intensification of long-haul flights and other highly polluting forms of tourist mobility. However, despite the manifest desire *to go back to where we were* that is reflected in these trends, post-pandemic tourism will unavoidably have to engage with the present challenges to global mobilities. These include the climate crisis, environmental crises, and migration crises (Tzanelli, 2022), all of which will place the 'politics of life' at the core of any attempt to mitigate or subvert existing forms of mobility injustice. In sum, the perspectives offered by biopolitical analyses may well prove key to determining whether the pandemic has taught us useful lessons for the future of tourism, or whether it has simply confirmed the increasingly exclusionary nature of modern tourism and its historical tendency to deny its own contribution to the very crises that beset it.

References

Brouder, P. (2020). Reset redux: Possible evolutionary pathways towards the transformation of tourism in a COVID-19 world. *Tourism Geographies, 22*(3), 484–490. https://doi.org/10.1080/14616688.2020.1760928

Brouder, S. T., Salazar, B., Mostafanezhad, M., Mei Pung, J., Lapointe, D., Higgins Desbiolles, F., Haywood, M., Hall, C. M., & Balslev Clausen, H. (2020). Reflections and discussions: Tourism matters in the new normal post COVID-19. *Tourism Geographies, 22*(3), 735–746. https://doi.org/10.1080/14616688.2020.1770325

Cresswell, T. (2021). Valuing mobility in a post COVID-19 world. *Mobilities, 16*(1), 51–65. https://doi.org/10.1080/17450101.2020.1863550

Dijstelbloem, H. O. (2021). *Borders as infrastructure: The technopolitics of border control.* The MIT Press.

Lew, A. A., Cheer, J. M., Haywood, M., Brouder, P., & Salazar, N. B. (2020). Visions of travel and tourism after the global COVID-19 transformation of 2020. *Tourism Geographies, 22*(3), 455–466. https://doi.org/10.1080/14616688.2020.1770326

Minca, C., & Roelofsen, M. (2023). Post-COVID biopolitical fantasies and the case of the Dutch "Pilot Holidays." *Environment and Planning C: Politics and Space.* https://doi.org/10.1177/23996544231194828a

Mostafanezhad, M., Cheer, J. M., & Sin, H. L. (2020). Geopolitical anxieties of tourism: (Im)mobilities of the COVID-19 pandemic. *Dialogues in Human Geography, 10*(2), 182–186. https://doi.org/10.1177/2043820620934206

Sheller, M. (2018). *Mobility justice: The politics of movement in an age of extremes.* Verso.

Sheller, M. (2021). Mobility justice and the return of tourism after the pandemic. *Mondes du tourisme,* https://journals.openedition.org/tourisme/3463

Sheller, M., & Urry, J. (2006). The new mobilities paradigm. *Environment and Planning a: Economy and Space, 38*(2), 207–226. https://doi.org/10.1068/a37268

Sheller, M., & Urry, J. (2016). Mobilizing the new mobilities paradigm. *Applied Mobilities, 1*(1), 10–25. https://doi.org/10.1080/23800127.2016.1151216

Tzanelli, R. (2022). Biopolitics in critical tourism theory: A radical critique of critique. *Via. Tourism Review, 21.* http://journals.openedition.org/viatourism/8242

Tzanelli, R. (2023a). Post-viral tourism's antagonistic tourist imaginaries. *Journal of Tourism Futures, 7*(3), 377–389. https://doi.org/10.1108/JTF-07-2020-0105

Tzanelli, R. (2023b). Economies of attention and the design of viable tourism futures. *Tourism Recreation Research, 48*(4), 605–615. https://doi.org/10.1080/02508281.2023.2188708

INDEX